齐康及其合作者建筑设计作品集

齐 康

中国建筑工业出版社

序

新中国成立的六十余年，特别是近三十年来，我和朋友、同事、学生们有机遇参与许多规划和建筑设计工作，使我和学生得到锻炼。实践是检验真理的标准，是提高理论的基础，理论反过来指导实践，环环上升。我们学习了国内外的优秀作品，懂得尊重历史的传承、外来文化的引入，最后达到创新和创意。我们探索行进。前辈们的素质和设计实践教育了我们。社会的进步和发展推动了我们。做学始终，能者为师，刻苦学习，自我启迪，团结和谐，与我的学生合作，这是一种团队精神。世界上最有价值的是事业，最珍贵的是友谊，最难得的是勤奋。我们十分重视进程、地区、层次、活动、对位、超前的哲学思辨，使设计作品上升为一种情感，并使之成为以人为本、持续发展的一种智慧结晶。我们在得到各级领导支持的同时也曾有过不少挫折、困惑，但困难面前总要坚强。这本书的出版要感谢我的合作者，感谢我的老师、学生、支持者，感谢我的家人的支持。人无完人，金无足赤，作品还有诸多遗憾。有信心地讲我们能行，同时谦逊地倾听读者的建议和意见。谢谢大家！

齐康

感谢我的老师杨廷宝先生，童寯先生！
感谢我的学生和合作者！
感谢我的家人多年来的支持！

Preface

Since the founding of our country, especially the recent thirty years, the development and progress of our society have driven our work. I have chance to paticipate and practice many works of architectural design and urban planning with my colleagues and students. We all gained much from the exercises. Practice as criterion of truth could improve the theory research, and vice versa. In these projects, we learned from other outstanding works in the relevant field both at home and abroad. We paid attention to come down local history and culture, and to draw in other cultures. We learned from elder architects' experiences. All efforts brought about innovation.

We fumbled for finding the right way to go on. We always remember to learn from person of ability, to study hard, to seek self-edification, to study from practice consistently and persistently, to promote solidarity and harmony. We considered philosophically problems of process, region, hierarchy, action, contraposition, and so on. We followed the idea of human oriented and sustainable development to make our works more mature and emotional. Of course we confronted many troubles in the process of work, but we overcame them and were supported by the local government. Career, friendship and hardworking are three most precious things for me. On the occasion of publishing this book, I want to thank my teathers, my students, my supportors and my family. Things always don't go as perfect as we think, we are willing to receive advices and suggestions from readers.

Qi Kang

Thanks to my teather Professor Yang, Professor Tong.
Thanks to my students and my colleagues.
Thanks to the perennial support of my family.

目录
Content

序
Preface

河南博物院 ··· 1
HENAN HISTORICAL MUSEUM

中国共产党代表团梅园新村纪念馆 ······················ 4
MEMORIAL OF MEIYUAN NEW VILLAGE

江苏淮安周恩来纪念馆 ······································· 6
ZHOU EN-LAI MEMORIAL IN HUAIAN

江苏淮安周恩来遗物陈列馆 ······························· 8
EXTENSION TO ZHOU EN-LAI MEMORIAL IN HUAIAN

江苏淮安周恩来纪念馆扩建工程 ······················· 11
ZHOU EN-LAI MEMORIAL EXPANSION PROJECT IN HUAIAN

侵华日军南京大屠杀遇难同胞纪念馆（一、二期）······ 13
MEMORIAL TO VICTIMS IN NANJING MASSACRE BY JAPANESE INVADERS

南京雨花台革命烈士纪念馆、碑 ······················· 16
THE MEMORIAL HALL AND THE MONUMENT OF THE YUHUATAI CEMETRY

南京雨花台警察纪念碑 ····································· 21
THE POLICEMEN MONUMENT OF YUHUATAI IN NANJING

福建武夷山庄 ··· 23
WU YI HOTEL, FUJIAN PROVINCE

福建武夷山老街（宋街） ································· 28
LAO STREET(SONG STREET), WUYI MOUNTAIN, FUJIAN PROVINCE

福建武夷山老街渔唱 ··· 33
YU CHANG LAO STREET, WU YI MOUNTAIN, FUJIAN PROVINCE

福建武夷山九曲宾馆 ··· 35
JIU QU HOTEL, WU YI MOUNTAIN, FUJIAN PROVINCE

福建武夷山玉女大酒店 ····································· 40
YUNV HOTEL, WUYI MOUNTAIN, FUJIAN PROVINCE

福建省武夷山碧丹酒家 ····································· 43
BIDAN HOTEL, WUYI MOUNTAIN, FUJIAN PROVINCE

福建武夷山幔亭山房 ··· 46
MANTING GUEST HOUSE, WUYI MOUNTAIN, FUJIAN PROVINCE

福建武夷山彭祖山房 ··· 50
PENGZU GUEST HOUSE, WUYI MOUNTAIN, FUJIAN PROVINCE

福建武夷山游船码头 ··· 51
THE PLEASURE-BOAT PIER, WUYI MOUNTAIN, FUJIAN PROVINCE

福建武夷山茶观 ··· 55
TEA CONCEPT, WUYI MOUNTAIN, FUJIAN PROVINCE

福建武夷山天心茶室 ··· 57
TIANXIN TEA HOUSE, WUYI MOUNTAIN, FUJIAN PROVINCE

福建武夷山天心亭 ·· 58
TIANXIN PAVILION, WUYI MOUNTAIN, FUJIAN PROVINCE

福建武夷山大王亭 ·· 59
DAWANG PAVILION, WUYI MOUNTAIN, FUJIAN PROVINCE

九一八历史博物馆 ·· 60
9.18 HISTORICAL MUSEUM

镇海口海防历史纪念馆 ····································· 62
COASTAL DEFENCE HISTORICAL MUSEUM, ZHENHAI, ZHEJIANG

中国人民解放军海军诞生地纪念馆 ···················· 68
NAVY MUSEUM OF CHINESE PEOPLE'S LIBERATION ARMY

江苏海安苏中七战七捷纪念碑、馆 ···················· 72
MEMORIAL OF THE SEVEN VICTORIES IN MID-JIANGSU, HAI'AN JIANGSU PROVINCE

淮海战役陈官庄地区歼灭战烈士陵园 ················ 74
HUAIHAI CAMPAIGN CHENGUANZHUANG ANNIHILATION REGION MARTYRS CEMETERY

黄山国际大酒店 ··· 77
INTERNATIONAL HOTEL OF HUANGSHAN

净月潭风景区整体设计 ····································· 79
HOLISTIC DESIGN OF JING YUE TAN SCENIC SPOT

福建长乐海螺塔 ··· 83
CONCH TOWER, CHANGLE, FUJIAN

浙江天台山济公佛院 ··· 85
JIGONG TEMPLE OF TIANTAI MOUNTAIN, ZHEJIANG PROVINCE

福建惠安海疆楼 ··· 90
HUIAN SEAFRONTIER BUILDING

中文	英文	页码
福建惠安海门亭	HUIAN HAIMEN PAVILION, FUJIAN PROVINCE	93
江阴望江楼	BELVEDERE IN JIANGYIN	94
西藏和平解放纪念碑	THE MONUMENT FOR THE PEACEFUL LIBERATION OF TIBET	96
福建省博物馆	MUSEUM OF FUJIAN PROVINCE	100
哈尔滨金上京历史博物馆	HISTORICAL MUSEUM OF JING CAPITAL, HARBIN	102
冰心文学馆	BINGXIN MEMORIAL	104
华罗庚纪念馆	HUA LUOGENG MEMORIAL	107
苏州丝绸博物馆	MUSEUM OF SILK IN SUZHOU	110
苏州丝绸博物馆改扩建	MUSEUM OF SILK EXPANSION DESIGN, SUZHOU	112
中国国学中心	CHINA NATIONAL SINOLOGY CENTER	113
南京1912民国文化街区	1912 CHINA CULTURAL DISTRICT, NANJING	115
中国鞋文化博物馆	THE CHINESE SHOES CULTURE MUSEUM, WENZHOU	120
大连贝壳博物馆	DALIAN CONCH MUSEUM, LIAONING PROVINCE	124
晨光1865科技创意产业园规划设计	CHENGUANG 1865 CREATIVITY SCIENCE TECHNOLOGY DESIGN, NANJING	126
晨光1865大门	CHENGUANG 1865 GATE, NANJING	127
晨光1865 VIP贵宾楼	CHENGUANG 1865 VIP BUILDING, NANJING	128
晨光1865东部入口区	EAST ENTRANCE ZONE OF CHENGUANG 1865, NANJING	131
南京中信大楼	ZHONGXIN TOWER, NANJING	133
南京达舜国际广场	DASHUN INTERNATIONAL SQUARE, NANJING	134
中国人民银行南京分行	NANJING BRANCH OF PEOPLE'S BANK OF CHINA	135
南京鼓楼邮政大楼	GULOU POST MANSION, NANJING	136
南京华侨大厦	OVERSEA CHINESE HOTEL, NANJING	138
南京怡华假日酒店	YIHUA HOTEL, NANJING	139
江苏省国税大厦	NATIONAL TAX BUILDING, JIANGSU PROVINCE	140
中国人民银行济南分行营业楼	JINAN BRANCH OF THE PEOPLE'S BANK OF CHINA	141
深圳国人大厦	GUOREN BUILDING OF SHENZHEN	142
南京古生物博物馆	MUSEUM OF PALEOBIOLOGY, NANJING	144
徐州兵马俑博物馆	TERRACOTTA WARRIORS AND HORSES MUSEUM, XUZHOU	148
淮北市博物馆	MUSEUM OF HUAIBEI CITY	150
长白山满族文化博物馆	CHANGBAI MOUTAIN MANCHU CULTURE MUSEUM	153
东莞东江纵队纪念馆	THE MEMORIAL HALL OF DONGJIANG COLUMN, DONGGUAN	155
东北沦陷史陈列馆	THE MUSEUM FOR THE HISTORY OF OCCUPATION IN NORTHEAST OF CHINA	156
吉林省集安市高句丽博物馆	KOGURYO HERITAGE MUSEUM IN JIAN, JILIN PROVINCE	157
安徽凤阳县小岗村"大包干"纪念馆	DABAOGAN MUSEUM, FENG YANG, ANHUI PROVINCE	159
东南大学九龙湖校区图书馆	LIBRARY OF JIULONGHU CAMPUS, SOUTHEAST UNIVERSITY	160
东南大学李文正科技楼	LI WENZHENG BUILDING, SOUTHEAST UNIVERSITY	162
东南大学榴园宾馆	LIUYUAN HOTEL OF SOUTHEAST UNIVERSITY	167
中国科学技术大学生命科学楼	LIFE SCIENCES BUILDING OF USTC	168
南京农业大学主楼	MAIN BUILDING OF NANJING AGRICULTURAL UNIVERSITY	169
南京农业大学逸夫教学楼	YIFU BUILDING OF NANJING AGRICULTURAL UNIVERSITY	170
南京农业大学金陵研究院	JINLING ACADEME OF NANJING AGRICULTURAL UNIVERSITY	171
南京农业大学第四教学楼	THE FOURTH ACADEMIC BUILDING OF NANJING AGRICULTURAL UNIVERSITY	173
大连理工大学伯川图书馆	BOCHUAN LIBRARY OF DALIAN UNIVERSITY OF TECHNOLOGY	174
青岛理工大学图书馆	LIBRARY OF QINGDAO TECHNOLOGICAL UNIVERSITY	177

中文	页码	英文
大连水产学院文夫图书馆	179	WENFU LIBRARY OF DALIAN FISHERIES COLLEGE
南京航空航天大学逸夫科技馆	180	YIFU BUILDING OF SCIENCE & TECHNOLOGY, NANJING UNIVERCITY OF AERONAUTICS
南京航空航天大学综合楼	181	COMPLEX BUILDING IN NANJING UNIVERSITY OF AERONAUTICS
中国矿业大学教学楼	182	MAIN BUILDING OF CHINA UNIVERSITY OF MINING AND TECHNOLOGY
徐州师范大学教学主楼群	183	MAIN BUILDINGS OF XUZHOU NORMAL UNIVERSITY
徐州师范大学艺术楼	185	THE ART BUILDING OF XUZHOU NORMAL UNIVERSITY
徐州工程学院教学主楼	187	MAIN BUILDING OF XUZHOU INSTITUTE OF TECHNOLOGY
徐州工程学院体育馆	190	THE GYMNASIUM OF XUZHOU INSTITUTE OF TECHNOLOGY
徐州工程学院图书馆	191	LIBRARY OF XUZHOU INSTITUTE OF TECHNOLOGY
盐城工学院主楼	192	MAIN BUILDING OF YANCHENG INSTITUTE OF TECHNOLOGY
南京鼓楼医院急救中心	193	EMERGENCY CENTER OF GULOU HOSPITAL, NANJING
南京鼓楼医院高层病房	196	HIGH-RISE WARD OF GULOU HOSPITAL, NANJING
河南许昌市政府大楼	198	CITY HALL OF XUCHANG, HENAN
江苏省常熟市政府大楼	199	CITY HALL OF CHANGSHU, JIANGSU
江苏省国家安全教育展览馆	201	NATIONAL SECURITY EDUCATION EXHIBITION HALL, JIANGSU
南京市中级人民法院审判庭	202	NANJING INTERMEDIATE PEOPLE'S COURT ADJUDICATION DIVISION
南京市栖霞区法院审判综合楼	203	COURT DESIGN OF XIXIA DISTRICT, NANJING
泉州东湖公园	204	DONGHU PARK, QUANZHOU
溧水永寿寺塔塔院	208	THE COURTYARD OF YONGSHOU PAGODA, LISHUI
镇江碧榆园	209	BIYU GARDEN OF ZHENJIANG
厦门鼓浪屿别墅宾馆	212	GULANGYU VILLA HOTEL OF XIAMEN
福建晋江八仙山公园	215	JINJIANG BAXIANSHAN PARK, FUJIAN
厦门园博园杏林阁	216	YUANBOYUAN XINGLIN PAVILION IN XIAMEN
张家界国家森林公园入口门票站	217	ZHANGJIAJIE NATIONAL FOREST PARK ENTRANCE TICKET STATION
江苏仪化大酒店	219	YIHUA HOTEL, JIANGSU
扬州西园大酒店	220	XIYUAN HOTEL, YANGZHOU
南京雨花台青少年活动中心	221	NANJING YUHUATAI CENTER FOR YOUTHS
江苏省钟山干部疗养院	223	ZHONGSHAN SANATORIUM OF JIANGSU PROVINCE
镇江市福利院	224	ZHENJIANG WELFARE HOSPITAL
中国科学院土壤研究所办公楼	225	BUILDING OF SOIL RESEARCH INSTITUTE, CHINESE ACADEMY OF SCIENCES
中国科学院澄江古生物研究站	226	CHENGJIANG PALEOBIOLOGY ACADEME, YUNNAN
黄山市电业局调度楼	227	ELECTRIC POWER SCHEDULING BUILDIND OF HUANGSHAN CITY
深圳贝岭居	228	BEILING BUILDING, SHENZHEN
厦门市文联大楼	229	CULTURE BUILDING OF XIAMEN
镇江市人防办防空指挥中心	230	COMMANDING CENTER OF CIVIL AERIAL DEFENCE IN ZHENJIANG
中山大学珠海分校伍舜德学术交流中心	231	WUSHUNDE SCIENTIFIC COMMUNICATION CENTER IN ZHONGSHAN UNIVERSITY
南京大厂中学	231	DACHANG MIDDLE SCHOOL, NANJING
扬州历史纪念碑	232	YANGZHOU HISTORIC MONUMENTS
邳州公园大门	233	GATE OF PEIZHOU GARDEN
南通烈士陵园纪念碑	233	NANTONG MARTYRS CEMETERY MONUMENTS
南京金陵中学一百周年纪念碑	234	THE MONUMENT FOR 100TH ANNIVERSARY JINLING SENIOR SCHOOL, NANJING
缙云县洋潭头后陈山公园景观塔	235	JINYUN YANGTANTOU LANDSCAPE TOWER IN HOUCHENSHAN PARK

江苏如皋红十四军纪念馆 ⋯⋯⋯⋯⋯⋯⋯⋯⋯⋯⋯⋯⋯⋯⋯ 237
RUGAO FOURTEEN RED ARMY MEMORIAL, JIANGSU

南昌八大山人纪念馆改扩建 ⋯⋯⋯⋯⋯⋯⋯⋯⋯⋯⋯⋯⋯ 238
BADASHANREN MEMORIAL HALL RENOVATION AND EXPANSION

淮北革命根据地纪念馆、纪念碑 ⋯⋯⋯⋯⋯⋯⋯⋯⋯⋯⋯ 239
HUAIBEI REVOLUTIONARY BASE MEMORIAL, MONUMENT

淮安八十二烈士纪念馆 ⋯⋯⋯⋯⋯⋯⋯⋯⋯⋯⋯⋯⋯⋯⋯ 240
EIGHTY-TWO MARTYRS MEMORIAL, HUAI'AN

淮安母爱公园爱心塔 ⋯⋯⋯⋯⋯⋯⋯⋯⋯⋯⋯⋯⋯⋯⋯⋯ 241
LOVE TOWER DESIGN, HUAI'AN

淮安感恩广场 ⋯⋯⋯⋯⋯⋯⋯⋯⋯⋯⋯⋯⋯⋯⋯⋯⋯⋯⋯ 242
ARCHITECTURAL DESIGN OF GAN'EN HALL, HUAI'AN

重庆杨闇公烈士陵园规划及建筑单体 ⋯⋯⋯⋯⋯⋯⋯⋯⋯ 243
YANG AN GONG MARTYRS CEMETERY, CHONGQING

南通市中小学生素质教育基地展览馆 ⋯⋯⋯⋯⋯⋯⋯⋯⋯ 244
PRIMARY AND MIDDLE SCHOOL-STUDENTS QUALITY EDUCATION EXHIBITION HALL IN NANTONG

厦门思明研发产业园 ⋯⋯⋯⋯⋯⋯⋯⋯⋯⋯⋯⋯⋯⋯⋯⋯ 246
SINGMING R & D INDUSTRIAL PARK IN XIAMEN

南通汽车客运东站 ⋯⋯⋯⋯⋯⋯⋯⋯⋯⋯⋯⋯⋯⋯⋯⋯⋯ 247
EASTERN AUTOMOBILE PASSENGER STSTION, NANTONG

姜堰博物馆 ⋯⋯⋯⋯⋯⋯⋯⋯⋯⋯⋯⋯⋯⋯⋯⋯⋯⋯⋯⋯ 250
MUSEUM, JIANGYAN

南京丁山宾馆 ⋯⋯⋯⋯⋯⋯⋯⋯⋯⋯⋯⋯⋯⋯⋯⋯⋯⋯⋯ 252
DINGSHAN HOTEL, NANJING

福州市马尾区图书馆 ⋯⋯⋯⋯⋯⋯⋯⋯⋯⋯⋯⋯⋯⋯⋯⋯ 253
THE LIBRARY IN MAWEI OF FUZHOU

江苏省金坛茅山气象综合实验基地 ⋯⋯⋯⋯⋯⋯⋯⋯⋯⋯ 254
METEOROLOGICAL COMPREHENSIVE EXPERIMENTAL BASE

宣城市宛陵湖配套商业服务建筑 ⋯⋯⋯⋯⋯⋯⋯⋯⋯⋯⋯ 256
COMMERCIAL SERVICES ARCHITECTURE IN XUANCHENG

南京大学东南楼 ⋯⋯⋯⋯⋯⋯⋯⋯⋯⋯⋯⋯⋯⋯⋯⋯⋯⋯ 257
SOUTHEAST BUILDING OF NANJING UNIVERSITY

泰州中学老校区图书馆 ⋯⋯⋯⋯⋯⋯⋯⋯⋯⋯⋯⋯⋯⋯⋯ 258
LIBRARY OF TAIZHOU HIGH SHOOL

南京五台山体育馆 ⋯⋯⋯⋯⋯⋯⋯⋯⋯⋯⋯⋯⋯⋯⋯⋯⋯ 259
WU TAI SHAN STADIUM, NANJING

锦溪人民医院（老年护理院）一期 ⋯⋯⋯⋯⋯⋯⋯⋯⋯⋯ 260
JINXI PEOPLE'S HOSPITAL, KUNSHAN

蚌埠规划勘测研究中心设计方案 ⋯⋯⋯⋯⋯⋯⋯⋯⋯⋯⋯ 261
ARCHITECTURE DESIGN FOR THE PLAN & SURVEY CENTER OF BENGBU

大连甘井子区图书档案馆设计方案 ⋯⋯⋯⋯⋯⋯⋯⋯⋯⋯ 261
GANJINGZI LIBRARY & ARCHIVES CENTER, DALIAN

禹州市钧窑遗址博物馆设计方案 ⋯⋯⋯⋯⋯⋯⋯⋯⋯⋯⋯ 262
YUZHOU JUN PORCELAIN MUSEUM DESIGN

青海湖旅游接待中心设计方案 ⋯⋯⋯⋯⋯⋯⋯⋯⋯⋯⋯⋯ 263
QINGHAI LAKE TOURISM RECEPTION CENTER DESIGN

微山县文化艺术中心设计方案 ⋯⋯⋯⋯⋯⋯⋯⋯⋯⋯⋯⋯ 264
CULTURAL ARTS CENTER DESIGN OF WEISHAN COUNTY

大连普兰店市青少年活动中心设计方案 ⋯⋯⋯⋯⋯⋯⋯⋯ 266
DESIGN OF PULANDIAN CENTER FOR YOUNG PEOPLE, DALIAN

青州博物馆设计方案 ⋯⋯⋯⋯⋯⋯⋯⋯⋯⋯⋯⋯⋯⋯⋯⋯ 267
QINGZHOU MUSEUM DESIGN

北京某区城市设计方案 ⋯⋯⋯⋯⋯⋯⋯⋯⋯⋯⋯⋯⋯⋯⋯ 268
URBAN PLANING OF BEIJING LOCAL DISTRICT

南京农业大学理科实验楼设计方案 ⋯⋯⋯⋯⋯⋯⋯⋯⋯⋯ 269
SCIENCE LABORATORY BUILDING OF NANJING AGRICULTURAL UNIVERSITY

北京市通州某地段规划设计方案 ⋯⋯⋯⋯⋯⋯⋯⋯⋯⋯⋯ 270
PLANNING AND DESIGN OF LOCAL TOWN IN TONGZHOU, BEIJING

镇江碧榆园国际会议中心设计方案 ⋯⋯⋯⋯⋯⋯⋯⋯⋯⋯ 271
ZHENJIANG BIYUYUAN INTERNATIONAL CONFERENCE CENTRE DESIGN

南京浦镇车辆有限公司1915公馆设计方案 ⋯⋯⋯⋯⋯⋯ 272
PUZHEN VEHICLES CO., LTD. 1915 RESIDENCE, NANJING

厦门市同安区佛心寺设计方案 ⋯⋯⋯⋯⋯⋯⋯⋯⋯⋯⋯⋯ 273
FOXIN TEMPLE IN TONGAN DISTRICT, XIAMEN

新一代天气雷达建设项目（天语舟）设计方案 ⋯⋯⋯⋯⋯ 274
A PROJECTION OF NEW GENERATION WEATHER RADAR CONSTRCTION (TIANYU SHIP)

义乌细菌战受难同胞纪念馆设计方案 ⋯⋯⋯⋯⋯⋯⋯⋯⋯ 275
MEMORIAL OF THE VICTIMS IN YIWU GERM WARFARE

北京亦庄环渤海高端总部基地规划设计方案 ⋯⋯⋯⋯⋯⋯ 276
YIZHUANG BOHAI HIGH-END HEADQUARTERS BLOCK PLANNING, BEIJING

南通大学附属医院新门诊楼设计方案 ⋯⋯⋯⋯⋯⋯⋯⋯⋯ 277
OUTPATIENT BUILDING OF AFFILIATED HOSPITAL OF NANTONG UNIVERSITY

天台安厦度假酒店设计方案 ⋯⋯⋯⋯⋯⋯⋯⋯⋯⋯⋯⋯⋯ 279
TIANTAI ANSHA RESORT HOTEL, TAIZHOU

南通东城涌鑫广场设计方案 ⋯⋯⋯⋯⋯⋯⋯⋯⋯⋯⋯⋯⋯ 281
DONGCHENG YONGXIN PLAZA, NANTONG

青岛大学科技研发中心设计方案 ⋯⋯⋯⋯⋯⋯⋯⋯⋯⋯⋯ 283
THE SCIENCE R & D CENTER OF QINGDAO UNIVERSITY

青岛大学医学教育综合楼设计方案 ⋯⋯⋯⋯⋯⋯⋯⋯⋯⋯ 284
MULTI-PURPOSED MEDICAL TEACHING BUILDING IN QINGDAO UNIVERSITY

武夷山"红山文化休闲（荣昌汇）"规划设计方案 ⋯⋯⋯ 285
PLANNING FOR RONGCHANGHUI IN WUYI MOUNTAIN

河南博物院
HENAN HISTORICAL MUSEUM

设 计 人：齐康、郑炘、王建国、 　　　　　张宏、李立、张彤、 　　　　　李秋海、刘爱华	Designer: QI Kang, ZHENG Xin, WANG Jian-guo, ZHANG Hong, LI Li, ZHANG Tong, LI Qiu-hai, LIU Ai-hua
工程地点：河南省郑州市	Location: Zhengzhou City, Henan Province
工程规模：78800 平方米	Total Area: 78,800 sq.m.
设计时间：1993–1998	Design Time: 1993-1998
建成时间：1998	Completion: 1998
合作单位：河南省建筑设计研究院	Co-operation: Henan Institute of Architectural Design
获奖情况：1999 年河南省优秀工程设计一等奖	Prize: The First Prize of Henan Provincial Architecture Academic Award, 1999

设计构思：

我们的立意构思突出以中原之气为核心，于是我们借鉴传统、针对地段特点采取有主有从的布局手法，取中外建筑审美意识的精华，运用现代技术和材料，集多种功能、简洁和谐为整体，最终使河南博物院成为一座雄伟、壮观、贯通古今、极富中原特色的现代博物院建筑群。

河南博物院整个形体呈金字塔形。围绕中央大厅，结合四个庭院簇拥着一组临时陈列展厅和序言大厅，犹如众心捧月，较好地处理了主体与陪衬的艺术关系。整个建筑群以主体建筑为中心，赋予各部门相同的风格和不同的功能，并以庭院、廊道有机的空间组合，使整个建筑显得主从分明、和谐统一。从总平面看，整个建筑群由九个体块组成，暗合了九鼎定中原的内涵。

总之，我们的设计力求线条简洁，整体造型壮丽浑厚、风格独特、气势恢宏、内涵丰富，既凝聚中原地区历史文化特色，又符合功能使用要求。

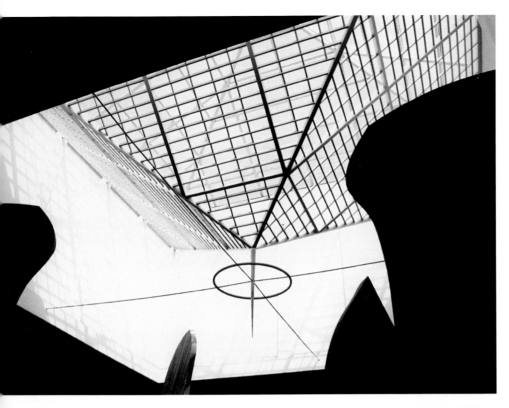

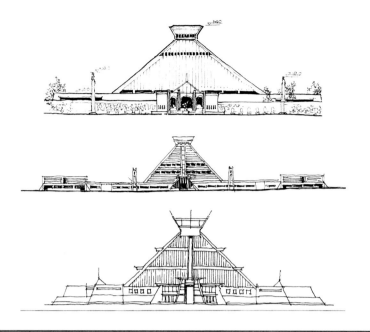
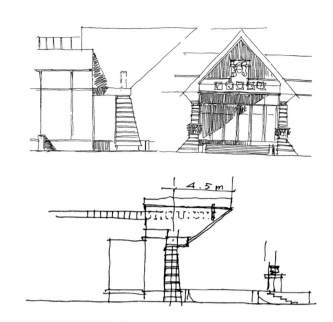

中国共产党代表团梅园新村纪念馆
MEMORIAL OF MEIYUAN NEW VILLAGE

设 计 人：齐康、许以立、孟建民、曹斌、仲德崑	Designer: QI Kang, XU Yi-li, MENG Jian-min, CAO Bin, ZHONG De-kun
工程地点：江苏省南京市	Location: Nanjing City, Jiangsu Province
工程规模：2400平方米	Total Area: 2,400 sq.m.
设计时间：1988–1990	Design Time: 1988-1990
建成时间：1991	Completion: 1991
合作单位：南京市建筑设计研究院	Co-operation: Nanjing Institute of Architectural Design
获奖情况：1992年国家优秀建筑设计金质奖	Prize: Gold Prize of Architecture Academic Award, 1992
1992年南京市优秀工程设计一等奖	The First Prize of Nanjing Architecture Academic Award,1992
1992年江苏省优秀工程设计一等奖	The First Prize of Jiangsu Provincial Architecture Academic Award, 1992
1992年建设部优秀工程设计一等奖	The First Prize of Architecture Academic Award, Construction Ministry,1992

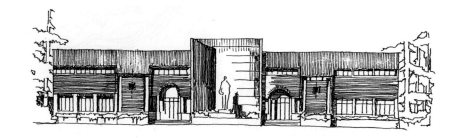

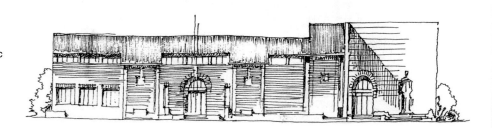

设计构思：

　　南京梅园是1945年以周恩来为首的中共代表团与国民党进行和平谈判时的住地。纪念馆从城市设计的角度出发，提出"两幢住宅"的概念，以"建筑环境的和谐，历史环境的再现"作为创作的中心思想，一方面将纪念馆融合到城市环境中，另一方面使观者触景生情，进而达到对历史事件的纪念。

　　纪念馆不自我突出，避免哗众取宠，以青灰面砖和黑色机瓦的二层坡顶建筑形象统一在里弄街坊的环境之中。宜人的比例，亲切的尺度，构成一座典雅的具有地方特色的时代建筑。内庭院中西院墙上部，以现代手法设计了一组透空窗，令人联想起当年国民党特务在钟岗里宿舍楼的"狼犬的眼睛"。内庭院迎面墙上是周总理生前喜爱的梅花做的窗饰。作为纪念馆中心的周恩来青铜塑像，出自周恩来从容步出梅园新村的历史照片。

　　梅园纪念馆，没有壮观的立面，没有名贵的材料，没有华丽的装饰，却有一种亲切的参与感。

江苏淮安周恩来纪念馆
ZHOU EN-LAI MEMORIAL IN HUAIAN

设 计 人：齐康、张宏、孟建民、张鹏举、齐昉	Designer: QI Kang, ZHANG Hong, MENG Jian-min, ZHANG Peng-ju, QI Fang
工程地点：江苏省淮安市	Location: Huaian City, Jiangsu Province
工程规模：3600 平方米	Total Area: 3,600 sq.m.
设计时间：1989–1992	Design Time: 1989-1992
建成时间：1992	Completion : 1992
合作单位：东南大学建筑设计研究院	Co-operation: Architects & Engineers Co., LTD of Southeast University
获奖情况：1993 年国家优秀建筑设计铜质奖 建设部优秀工程设计三等奖 江苏省优秀工程设计一等奖	Prize: The Brouze Medal of National Architecture Design Award, 1993 The Third Prize of Architecture Academic Award, Construction Ministry The First Prize of Jiangsu Provincial Architecture Academic Award

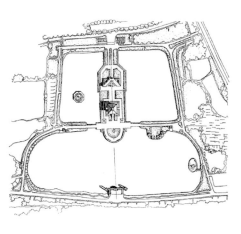

设计构思：

纪念馆区总体布局充分利用了自然水面和水中陆地，组织纪念性空间序列，创造出水天一色的视觉效果，体现出周总理伟大、光辉、亲切的形象和博大的胸怀。

纪念馆将简洁的锥形屋盖、方台体的基座、八角棱柱形的主体、4 根 11 米高的方柱、三角形的大台阶这些基本形体抽象地组合在一起，表现了一种扭转乾坤的力度，在形式感上追求建筑的地方性、民族精神以及"永恒"这一主题的表达。

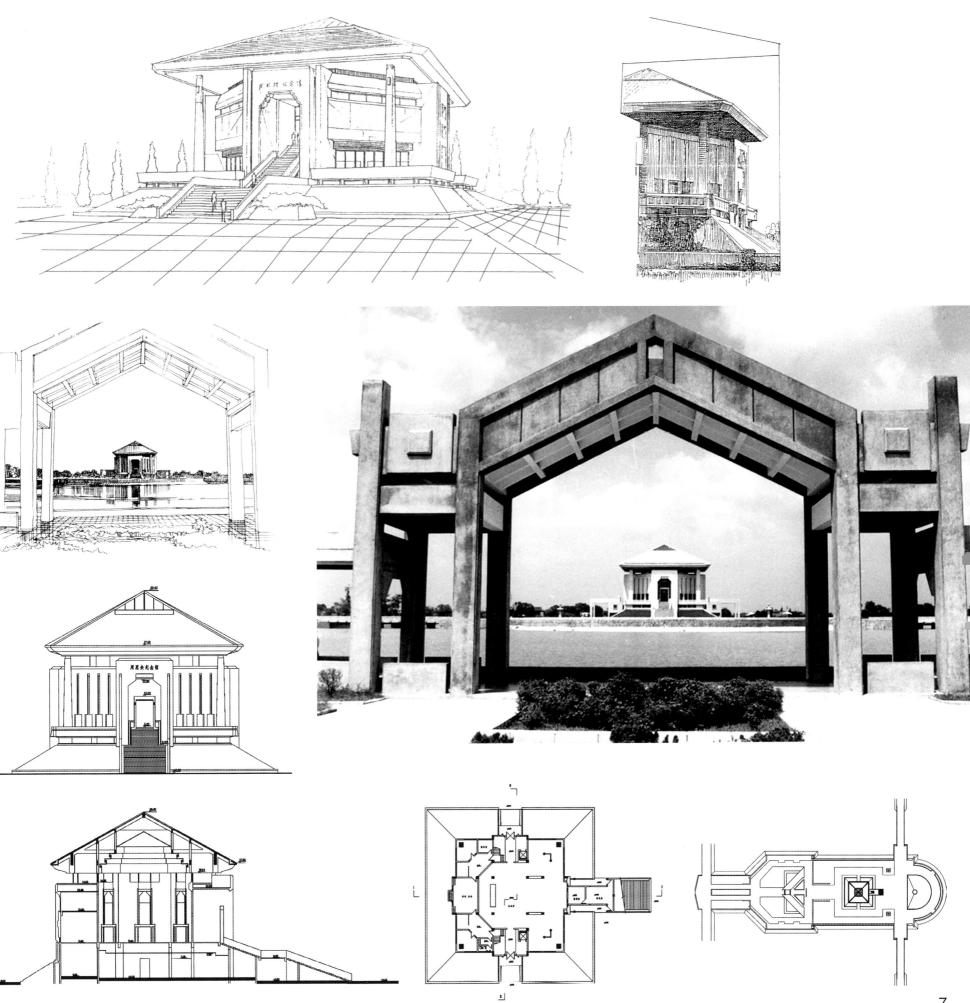

江苏淮安周恩来遗物陈列馆
EXTENSION TO ZHOU EN-LAI MEMORIAL IN HUAIAN

设 计 人：齐康、张宏、于雷
工程地点：江苏省淮安市
工程规模：3820平方米
设计时间：1996.1–1996.5
建成时间：1997.10
合作单位：东南大学建筑设计研究院
获奖情况：2000年教育部优秀工程设计二等奖

Designer:　　QI Kang, ZHANG Hong, YU Lei
Location:　　Huaian City, Jiangsu Province
Total Area:　 3,820 sq.m.
Design Time: January, 1996-May, 1996
Completion:　October, 1997
Co-operation: Architects & Engineers Co., LTD of Southeast University
Prize:　　　　The Second Prize of Architecture Academic Award, Education Ministry, 2000

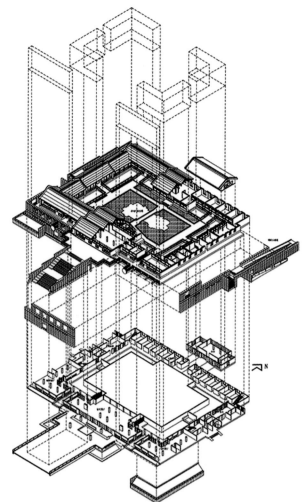

设计构思：

周恩来遗物陈列馆是周恩来纪念馆的扩建工程，其基地位于纪念馆轴线的北端。设计者面临两个难题，一是如何使原有轴线上的建筑在空间序列向北延伸后达到"转、承、起、合"的要求；另一方面则是如何使轴线尽端的四合院成为整个纪念建筑群的有机组成部分。

新的建筑群体由室外总理铜像、广场、桥、新"西花厅"组成，与整个周围的人工和自然环境求得新的协调。我们将穿过配楼两边长有高大水杉的中央通道作为一种封闭空间的过渡，将原来作为结束点的"伟亭"搬移到纪念馆前圆形环堤的南向顶点，作为前部轴线的空间程序的一个连接点，而新的室外总理铜像成为新延续序列的起点，并作为整个空间过渡。从两侧绕过铜像，就可以看到遗物陈列馆的总体形象。

对于遗物陈列馆我们设计时将它抬高一层，通过桥后进入馆底层的展览厅。通过大台阶可上至二层西花厅，人们可以瞻仰总理生前的住所和接待外来客人的前厅。底层的四合院内环四周的水池利用水位的高差，在出口处形成水瀑布与文渠沟通，既丰富了群体，又产生了动感。

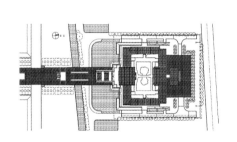

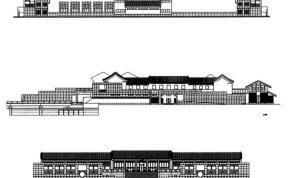

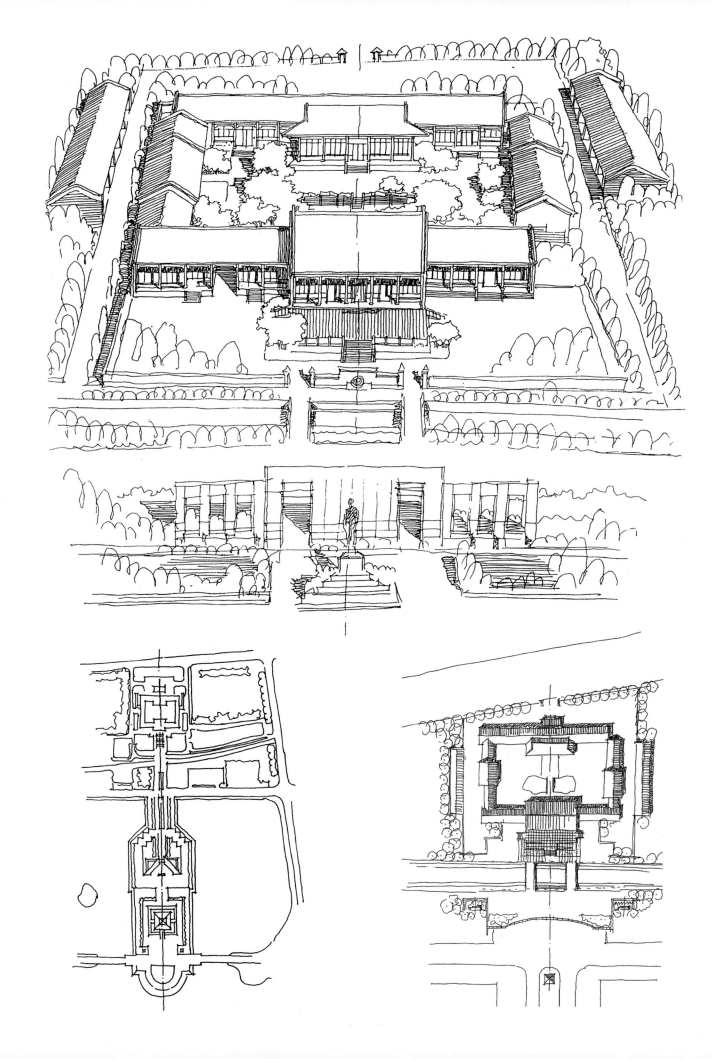

江苏淮安周恩来纪念馆扩建工程
ZHOU EN-LAI MEMORIAL EXPANSION PROJECT IN HUAIAN

设 计 人：齐康、叶菁
工程地点：江苏省淮安市
工程规模：3818 平方米
设计时间：2006
建成时间：2008
合作单位：东南大学建筑设计研究院

Designer:　　QI Kang, YE Jing
Location:　　Huaian City, Jiangsu Province
Total Area:　　3,818 sq.m.
Design Time:　2006
Completion:　 2008
Co-operation:　Architects & Engineers Co., LTD
　　　　　　　of Southeast University

侵华日军南京大屠杀遇难同胞纪念馆（一、二期）
MEMORIAL TO VICTIMS IN NANJING MASSACRE BY JAPANESE INVADERS

设 计 人：齐康、顾国强、郑嘉宁、张宏、寿刚、朱雷
工程地点：江苏省南京市
工程规模：4000平方米（一期），400平方米（二期）
设计时间：1983–1985（一期），1995（二期）
建成时间：1995（一期），1997（二期）
合作单位：南京市建筑设计研究院
获奖情况：1988年江苏省优秀建筑设计二等奖
　　　　　国家20世纪80年代优秀建筑创作十大作品第二名
　　　　　20世纪80年代国家优秀环境艺术一等奖

Designer:	QI Kang, Gu Guo-qiang, ZHENG Jia-ning, ZHANG Hong, SHOU Gang, ZHU Lei
Location:	Nanjing City, Jiangsu Province
Tatal Area:	4,400 sq.m.
Design Time:	1983-1985, 1995
Completion:	1995-1997
Co-operation:	Nanjing Institute of Architectural Design
Prize:	The Second Prize of Jiangsu Provincial Archi-tecture Academic Award, 1988 The Second Prize of Top Ten Architectural Design, 1980's The First Prize of National Environment Art, 1980's

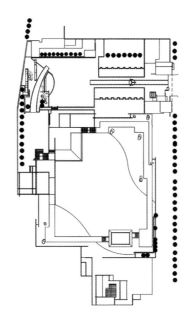
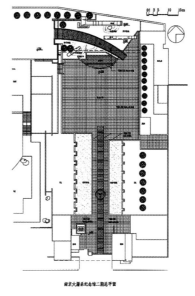
南京大屠杀纪念馆二期总平面

设计构思：

为纪念抗日战争胜利40周年，南京市政府决定建设纪念馆。地点设在南京当年掩埋死难者的十三个场地之一——江东门。建筑的特色在于用环境来表达纪念，创造一种意义的场所表现方式。纪念墙镌刻着醒目的中、英、日三国文字表述遇难者30万，由上而下，俯览全景。卵石广场象征死亡，与周边的青草构成一种生与死的对比。2米高、50米长的错落的围墙上，铭刻了当年屠杀的情景。卵石广场上的枯树，也是一种对日军"烧光、抢光、杀光"的暗示，塑造出十分悲剧性的环境。

与纪念馆呈对角的是尸骨陈列室，陈列了从地下挖出的死难者的尸骨，由地坪而下，表达了进入掩埋状态的概念。走出陈列室向上登上台阶就又回到悲惨的场景。沿着环绕的参观路线，布置了13块纪念石，每块代表一处在南京的掩埋地。当我们进入纪念展览时，可以感受到一种进入墓室之感，一种墓冢的象征。高低错落的石墙面、寸草不生的卵石广场、枯树、形似墓冢的纪念馆、近50米长的浮雕，营造出沉闷、悲愤、荒凉和凄凉交织的场景，以强烈的纪念氛围激起人们正义、善良的感情，探求世界永久的和平，寻求人们走向共同的友谊。第二期，设计者和管理者计划在尸骨馆对面，砌筑一面"哭墙"，刻上死难者的名字。它既是一块纪念死者的墙，又是一座碑，中间的细缝给人一种"劈开"的概念，下部摆上一个简朴的花圈，一种无声的纪念和有声的哀悼交融在一起。并在入口的场地上竖了一块纪念标志碑，刻上"1937.12——1938.1"。回顾历史，铭记"前事不忘，后事之师。"

1996.9.14 思想之灵泥! 1:200

南京雨花台革命烈士纪念馆、碑
THE MEMORIAL HALL AND THE MONUMENT OF THE YUHUATAI CEMETRY

设 计 人：杨廷宝、齐康、陈家骅、郑炘、孟建民、
　　　　　杨永龄、纪惠诛、李恕、曹斌等
工程地点：江苏省南京市
工程规模：5900平方米
设计时间：1983—1986
建成时间：1986
合作单位：南京市建筑设计研究院
获奖情况：1992年国家优秀建筑设计铜质奖
　　　　　建设部优秀工程二等奖
　　　　　江苏省优秀工程一等奖

Designer: YANG Ting-bao, QI Kang, CHEN Jia-bao,
 ZHENG Xin, MENG Jian-min, YANG Yong-ling,
 JI Hui-zhu, LI Shu, CAO Bin
Location: Nanjing City, Jiangsu Province
Total Area: 5900 sq.m.
Design Time: 1983-1986
Completion: 1986
Co-operation: Nanjing Institute of Architecural Design
Prize: The Bronze Medal of National
 Architecture Design Award, 1992
 The Second Prize of Architecture
 Academic Award, Construction Ministry
 The First Prize of Jiangsu Provincial
 Architecture Academic Award

设计构思：

　　雨花台坐落在南京城南。这里是著名的历史名胜区，有知名的云光师讲经台、方孝孺墓等，大革命时期和国内革命战争年代这里成了国民党反动势力残杀共产党人和革命人士的场地，先后有十余万革命者在此就义。

　　新中国成立后，曾修建临时性纪念碑。20 世纪 80 年代初，在杨廷宝先生的主持下，以全国设计竞赛为基础，经过长期与专家们的研讨，形成了初步的构想，1982 年开始正式实施。以"轴线"统一了自然山丘与建筑群，通过建筑与自然的围合、建筑的围合、半人工围合，直到最终空间的开敞，渐次达到空间序列的高潮。从忠魂亭到北殉难处，整个中轴线长达 1000 余米，是我国现代纪念建筑中最长的轴线。

　　传统历史形成的空间及其氛围，是标志地区风格特征之一。中国古代建筑群体的轴线是依据一层层、一进进的空间围合形成的，它在层次上有序列、有主从、有层次。这种特征是东方建筑重要的组成。"轴线"在人们意识观念中存在着强烈的印象。建筑轴线有对称和不对称，有意念的、精神的、虚的，也有实体的。在新的环境氛围，寻求新的概念，这是一种探求。建筑群体依照杨廷宝先生定下的带有传统特色的现代建筑形式，对传统建筑形式加以变化，以简洁的手法表达传统建筑精神。利用半人工围合以及空间的开敞和封闭，达到空间序列的高潮，这是构思的主题。

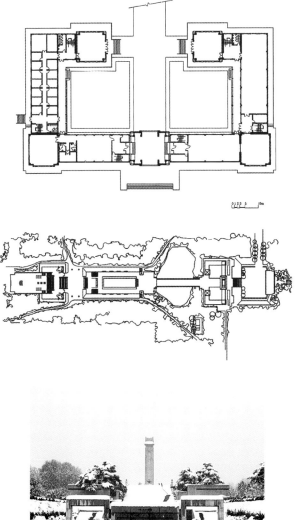

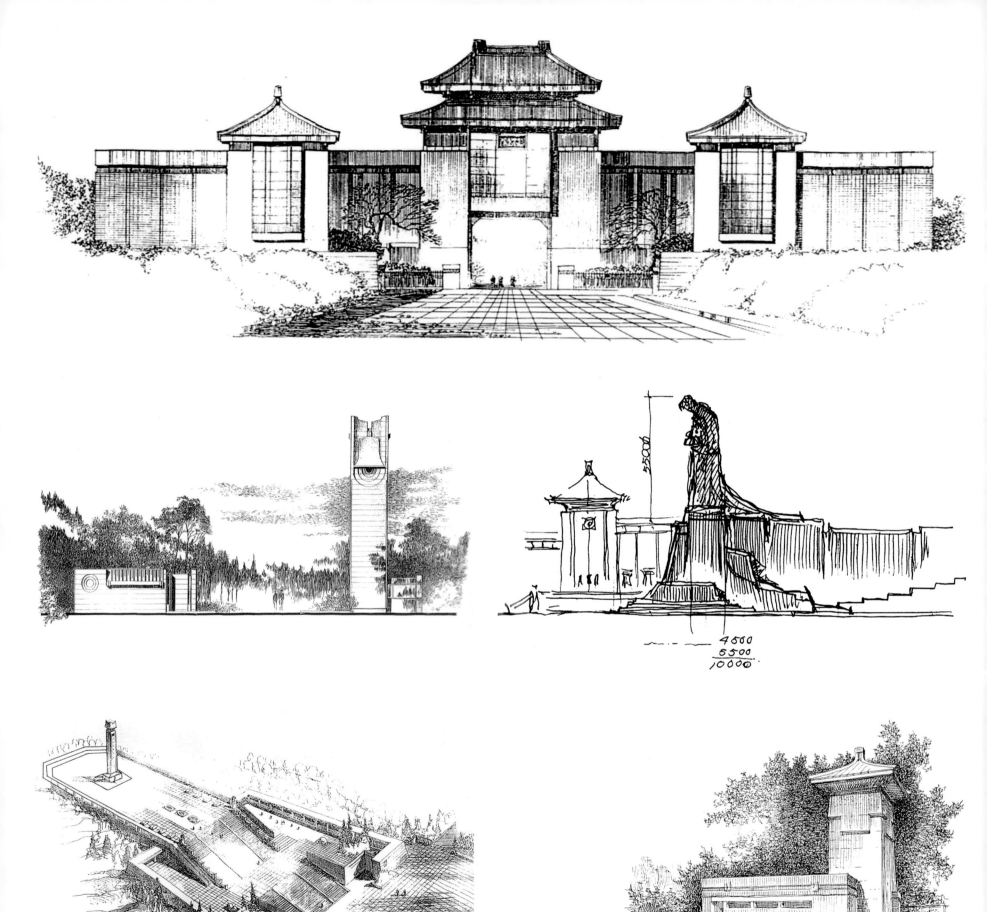

南京雨花台警察纪念碑
THE POLICEMEN MONUMENT OF YUHUATAI IN NANJING

设 计 人：齐康、王彦辉、张四维、钱大泾
工程地点：江苏省南京市
设计时间：2007
建成时间：2008.3

Designer: QI Kang, WANG Yan-hui,
　　　　　ZHANG Si-wei, QIAN Da-jing
Location: 　Nanjing City, Jiangsu Province
Design Time: 2007
Completion: 　March, 2008

福建武夷山庄
WU YI HOTEL, FUJIAN PROVINCE

设 计 人：	齐康、赖聚奎、杨子伸、陈宗钦、杨德安、蔡冠丽
工程地点：	福建省武夷山
工程规模：	2580平方米
设计时间：	1982–1983
建成时间：	1983
合作单位：	福建省建筑设计研究院
获奖情况：	1984年国家设计一等奖 建设部优秀工程设计一等奖 福建省优秀工程一等奖 国家20世纪80年代优秀建筑创作十大作品第三名

Designer:	QI Kang, LAI Ju-kui, YANG Zi-shen, CHEN Zong-qin, YANG De-an, CAI Guan-li
Location:	Wuyi Mountain, Fujian Province
Total Area:	2,580 sq. m.
Design Time:	1982-1983
Completion:	1983
Co-operation:	Fujian Institute of Architectural Design
Prize:	The First Prize of National Design Award, 1984 The First Prize of Architecture Academic Award, Construction Ministry The First Prize of Fujian Provincial Architecture Academic Award The Third Prize of Top Ten Architectural Design, 1980's

23

设计构思：

这是一组旅馆建筑群，此作品颇负盛名。当时曾多次选址，最后定在大王峰一麓、面对崇阳溪的斜坡地上。设计之初曾设想多层和低层二种。最后我们采用了错落有致的形体，沿着斜坡南向，平行布置，使之具有最好的风景面，把最多的自然风景引入到群组空间来。整个设计以"宜低不宜高，宜散不宜聚，宜土不宜洋"为原则，采取整体规划、分期实施、逐步调整的步骤。

由于行列布置，群体的垂直线对组合建筑群有重要的意义，需要尽可能地留出发展余地。配套的设备用房、办公部分在草坪前的半地下，得体而不损害坡地地形。入口在公路边，沿斜坡进入。建筑设计植根于地方风情文化，对地方的新风格进行了探索：斜坡顶，出挑垂莲柱的檐口和八角形廊边小窗。

整个建筑融汇于风景环境之中，高低起伏，顺应地势，叠瓦穿檐，自由错落，创造出既有浓郁乡土气息，又有明显时代特征的地方建筑新风格。内庭组合院落，运用奇特的山石，有野趣；门厅后的平台，利用山石筑成石凳、石桌，草坪、树木将一座红色屋顶、白色墙面组成的群体融合一气。室内外环境相得益彰，相辅成景。它背靠大王峰，山峦气势压人，庄丽而雄伟，山峰拔地 400 米，极其壮观。外墙的木构架装饰更丰富了形体的变化。随着二期工程的扩建，水池瀑布，洒下宴会厅的窗帘。浓缩的山涧、水溪，侧厅的室内布置别具风格，置身其中，宛若进入仙境一样。

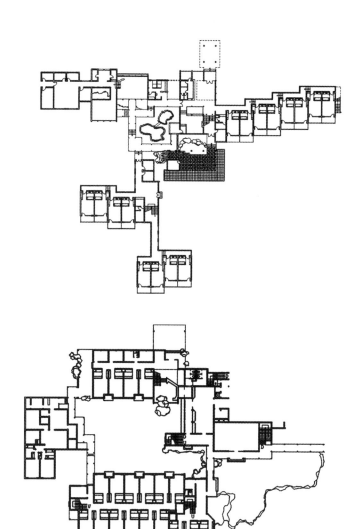

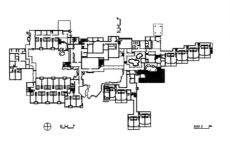

福建武夷山老街（宋街）
LAO STREET(SONG STREET), WUYI MOUNTAIN, FUJIAN PROVINCE

设 计 人：齐康、赖聚奎、陈宗钦、
　　　　　杨德安、蔡冠丽、刘叙杰
工程地点：福建省武夷山
设计时间：1986
建成时间：1988
合作单位：福建省建筑设计研究院

Designer:　QI Kang, LAI Ju-kui, CHEN Zong-qin,
　　　　　Yang De-an, CAI Guan-li, LIU Xu-jie
Location:　Wuyi Mountain, Fujian Province
Design Time: 1986
Completion:　1988
Co-operation: Fujian Institute of Architectural Design

福建武夷山老街渔唱
YU CHANG LAO STREET, WU YI MOUNTAIN, FUJIAN PROVINCE

设 计 人：陈宗钦		Designer:	CHEN Zong-qin
工程地点：福建省武夷山		Location:	Wuyi Mountain, Fujian Province
工程规模：30 平方米		Total Area:	30 sq.m.
设计时间：1988		Design Time:	1988
建成时间：1989		Completion:	1989
合作单位：福建省建筑设计研究院		Co-operation:	Fujian Institute of Architectural Design

福建武夷山九曲宾馆

JIU QU HOTEL, WU YI MOUNTAIN, FUJIAN PROVINCE

设 计 人：齐康、张宏、陈继良
工程地点：福建省武夷山
工程规模：4600平方米
设计时间：1991–1992
建成时间：1992
合作单位：东南大学建筑设计研究院

Designer:	QI Kang, ZHANG Hong, CHEN Ji-liang
Location:	Wuyi Mountain, Fujian Province
Total Area:	4,600 sq.m.
Design Time:	1991-1992
Completion:	1992
Co-operation:	Architectural Design & Research Institute of Southeast University

福建武夷山玉女大酒店
YUNV HOTEL, WUYI MOUNTAIN, FUJIAN PROVINCE

设 计 人：陈宗钦、齐康、段进、周明、张宏等
工程地点：福建省武夷山
工程规模：23000 平方米
设计时间：1987–1993
建成时间：1993
合作单位：深圳市蛇口工业区设计公司

Designer:	CHEN Zong-qin, QI Kang, DUAN Jin, ZHOU Ming, ZHANG Hong, etc
Location:	Wuyi Mountain, Fujian Province
Total Area:	23,000 sq.m.
Design Time:	1987-1993
Completion:	1993
Co-operation:	Shenzhen, Shekou Industrial Zone Design Company

福建武夷山幔亭山房
MANTING GUEST HOUSE, WUYI MOUNTAIN, FUJIAN PROVINCE

设 计 人：齐康、赖聚奎、陈宗钦、杨子伸、陈建霖	Designer: QI Kang, LAI Ju-kui, CHEN Zong-qin, YANG Zi-shen, CHEN Jian-Lin
工程地点：福建省武夷山	Location: Wuyi Mountain, Fujian Province
工程规模：1120平方米	Total Area: 1,120 sq.m.
设计时间：1980	Design Time: 1980
建成时间：1981	Completion: 1981
合作单位：福建省建筑设计研究院	Co-operation: Fujian Institute of Architectural Design

福建武夷山彭祖山房
PENGZU GUEST HOUSE, WUYI MOUNTAIN, FUJIAN PROVINCE

设 计 人：蔡冠丽
工程地点：福建省武夷山
设计时间：1983
建成时间：1984
合作单位：福建省建筑设计研究院

Designer: 　CAI Guan-Li
Location: 　Wuyi Mountain, Fujian Province
Design Time: 1983
Completion: 　1984
Co-operation: Fujian Institute of Architectural Design

福建武夷山游船码头
THE PLEASURE-BOAT PIER, WUYI MOUNTAIN, FUJIAN PROVINCE

设 计 人：赖聚奎、齐康
工程地点：福建省武夷山
工程规模：320平方米
设计时间：1980
建成时间：1981
合作单位：福建崇安建筑设计室

Designer: LAI Ju-kui, QI Kang
Location: Wuyi Mountain, Fujian Province
Total Area: 320 sq.m.
Design Time: 1980
Completion: 1981
Co-operation: Fujian Chongan Institute of Architectural Design

福建武夷山茶观
TEA CONCEPT, WUYI MOUNTAIN, FUJIAN PROVINCE

设 计 人：陈宗钦
工程地点：福建省武夷山
设计时间：1981
建成时间：1982

Designer: CHEN Zong-qin
Location: Wuyi Mountain, Fujian Province
Design Time: 1981
Completion: 1982

福建武夷山天心茶室
TIANXIN TEA HOUSE, WUYI MOUNTAIN, FUJIAN PROVINCE

设 计 人：齐康、晏隆余
工程地点：福建省武夷山
工程规模：40平方米
设计时间：1980
建成时间：1981

Designer: QI Kang, YAN Long-yu
Location: Wuyi Mountain, Fujian Province
Total Area: 40 sq.m.
Design Time: 1980
Completion: 1981

福建武夷山天心亭
TIANXIN PAVILION, WUYI MOUNTAIN, FUJIAN PROVINCE

设 计 人：齐康
工程地点：福建省武夷山
设计时间：1980
建成时间：1981

Designer:　　 QI Kang
Location:　　 Wuyi Mountain,
　　　　　　 Fujian Province
Design Time: 1980
Completion:　1981

福建武夷山大王亭
DAWANG PAVILION, WUYI MOUNTAIN,
FUJIAN PROVINCE

设 计 人：齐康
工程地点：福建省武夷山
设计时间：1980
建成时间：1981

Designer: QI Kang
Location: Wuyi Mountain, Fujian Province
Design Time: 1980
Completion: 1981

九一八历史博物馆
9.18 HISTORICAL MUSEUM

设 计 人：齐康、金俊、王彦辉、孙子磊 工程地点：辽宁省沈阳市 工程规模：12600平方米 设计时间：1997.9–1999.8 建成时间：1999.9 合作单位：中建中国东北建筑设计研究院 获奖情况：2000年辽宁省优秀工程设计一等奖 　　　　　2000年中国建筑工程总公司直属设计院优秀工程设计一等奖	Designer: QI Kang, JIN Jun, WANG Yan-hui, SUN Zi-lei Location: Shenyang City, Liaoning Province Total Area: 12,600 sq.m. Design Time: September, 1997-August, 1999 Completion: September, 1999 Co-operation: China Northeast Architectural Design and Research Institute Prize: The First Prize of Provincial Academic Award of Architecture, Liaoning Province, 2000 　　　　The First Prize of Design Award of the Design Institutes directly under China Architecture and Building Corporation	设计构思： 　　博物馆设在沈阳九一八残历碑的背后，是一块狭长的用地，长约700米，宽约45米左右，一侧是铁路干线，另一侧是城市干道。长达400米的建筑成为道路的一个重要界面。 　　尊重历史的残历碑是我们规划设计的首要构思。新建的博物馆用延伸的墙体环抱着它。纵长的围墙，既挡住铁路上过往火车的噪声，又是广场的界面。防噪声是我们的重要构思，我们设想采用单向坡来解决，使车轨摩擦的噪声有利于沿着斜坡屋顶而减小。 　　场地位于丁字路口，是观赏的对景，这有利于将这一纵长的建筑作为视景的结束点，因此在"残历碑"后的广场通向建筑的结束处设立了一座呈"Y"形的抗战纪念碑，碑高28米。它既是道路的对景，又表达展馆内抗战的一种象征性的结束，一种胜利的纪念。

镇海口海防历史纪念馆
COASTAL DEFENCE HISTORICAL MUSEUM, ZHENHAI, ZHEJIANG

设 计 人：齐康、张彤、段华朴
工程地点：浙江省宁波市
工程规模：3200平方米
设计时间：1994.3–1997.3
建成时间：1997.6
合作单位：浙江省建筑设计研究院
获奖情况：1998年建设部优秀工程设计三等奖
　　　　　"浙江杯"优秀工程设计金质奖

Designer:	QI Kang, ZHANG Tong, DUAN Hua-pu
Location:	Ningbo City, Zhejiang Province
Total Area:	3,200 sq.m.
Design Time:	March, 1994-March, 1997
Completion:	June, 1997
Co-operation:	Zhejiang Institute of Architectural Design
Prizes:	The Third Prize of National Academic Award of Architecture Design, Construction Ministry, 1998
	The Gold Prize of Provincial Academic Award of Architectural Design, Zhejiang Province

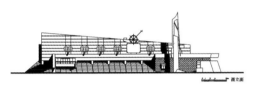

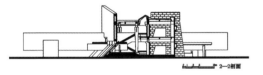
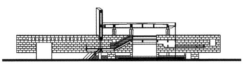

设计构思：

镇海口海防历史纪念馆位于浙江省宁波市镇海区。镇海口东屏舟山群岛，控扼甬江出海口，蛟门虎踞，形势险要，自古兵家必争。从明代以来，这里经历的抗击外来侵略的大小战役达46次之多。

从构思的最初，把潜在的场所精神集结于建筑，就成为设计的主旨。这种场所精神蕴涵在500年海防历史沉淀下来的历史记忆和精神特质中。

纪念馆的形体和空间被分成东西两个相互对比的部分。在建筑西部，面向招宝山公园，一道长达60米、厚1.5米的石墙从北到南斜插入建筑形体。它顺延了招宝山的形势，将山的尺度引入建筑之中。上部竖直、下部倾斜的形式来自不远处城塘合一的塘堤，体现出海防构筑强烈的抵御感。在竖直和倾斜之间，一道微弱而神秘的光线穿透石墙，贯穿在背后高通两层的空间之中。

在建筑的东部是上下两组各四个展厅。6.6米×7.2米的功能性体块被开启、打破、分散和重构。分隔展厅的四片石墙冲出屋面和外墙，成为建筑东立面具有强烈节奏感的控制性要素。这些穿插在内外之间的墙体，其体量是独立的，材料、色彩和质感在内外空间中也是一致的。实体之间的交接处让出一道缝隙，光和空间在这些墙体之间流动。建筑的品质和秩序延伸至室外，分离在更大的范围内创造了连续。

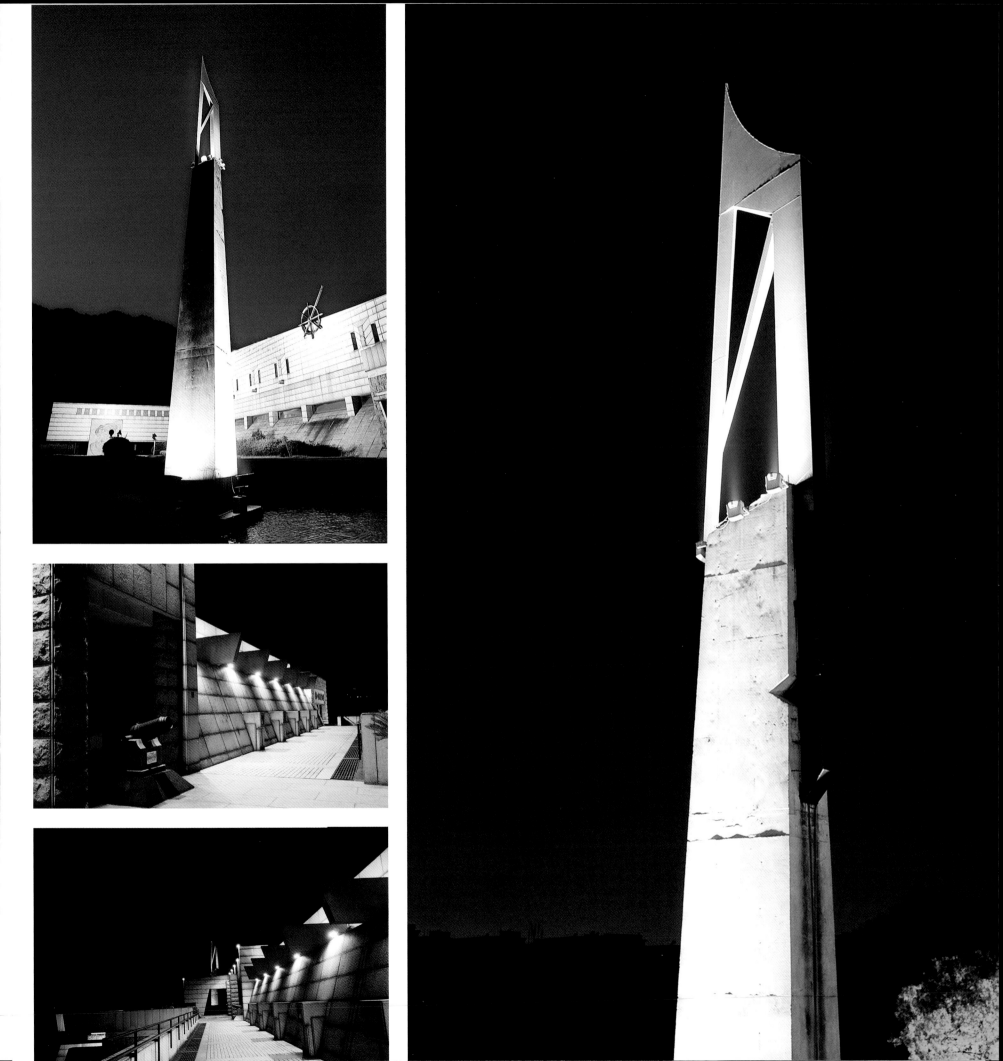

中国人民解放军海军诞生地纪念馆
NAVY MUSEUM OF CHINESE PEOPLE'S LIBERATION ARMY

设 计 人：齐康、张彤、黄印武、邹式汀	Designer: QI Kang, ZHANG Tong, HUANG Yin-wu, ZOU Shi-ding
工程地点：江苏省泰州市	Location: Taizhou City, Jiangsu Province
工程规模：2940平方米	Total Area: 2,940 sq.m.
设计时间：1998.9–1999.4	Design Time: September, 1998-April, 1999
建成时间：1999.4	Completion: April, 1999
合作单位：南京市建筑设计研究院	Co-operation: Nanjing Institute of Architectural Design
获奖情况：2000年建设部优秀工程设计二等奖	Prize: The Second Prize of Architecture Academic Award, Construction Ministry, 2000
2000年江苏省优秀工程设计一等奖	The First Prize of Jiangsu Provincial Architecture Academic Award, 2000
2000年南京市优秀工程设计一等奖	The First Prize of Nanjing Architecture Academic Award, 2000

设计构思：

15个8m×8m的方块，两层叠合，可以提供大约1800平方米的展厅面积，它们构成了纪念馆的主体，是所有形式生成和转化的基本。报告厅和办公储藏等用房被单独拉开，成为一个集中的独立体量，用二层的连廊与主体建筑相接。

最初的转化是将位于中间偏后的一个方块旋转15°，并脱离网格。转动的方块围合出一个特殊的空间，它贯通两层，拔出屋面。在以后的设计中，一个圆筒替代这个转动的方块，保留了特殊，模糊了方向。

这是整个形式系统中空无的中心。方块在圆筒的顶部复又出现，光从方和圆的间隙中洒落下来，使建筑的中心变得明亮而神秘。在圆弧形墙面和玻璃砖构筑的方形格网界面之间是连接上、下展厅的主要楼梯。一道7.25米高的花岗石墙体从中心直接冲出室外，在它的一侧是长达44.66米的坡道，它们切开方块组成的矩阵，把参观者从室外直接邀请进建筑中心的虚无。

在这个特殊的表皮和规整的展厅之间，同样是贯通两层的空间。一系列白色混凝土薄片支撑起屋顶天窗。由于这些薄片的两边分别搭接在曲面和直面上，它们在透视方向上形成了一种转动的渐变，将光线格构出优美的韵律。

在这里，光和虚无再一次成为整形和特异之间的主题，是分离出特异，赋予形式转化以自然的诗意。

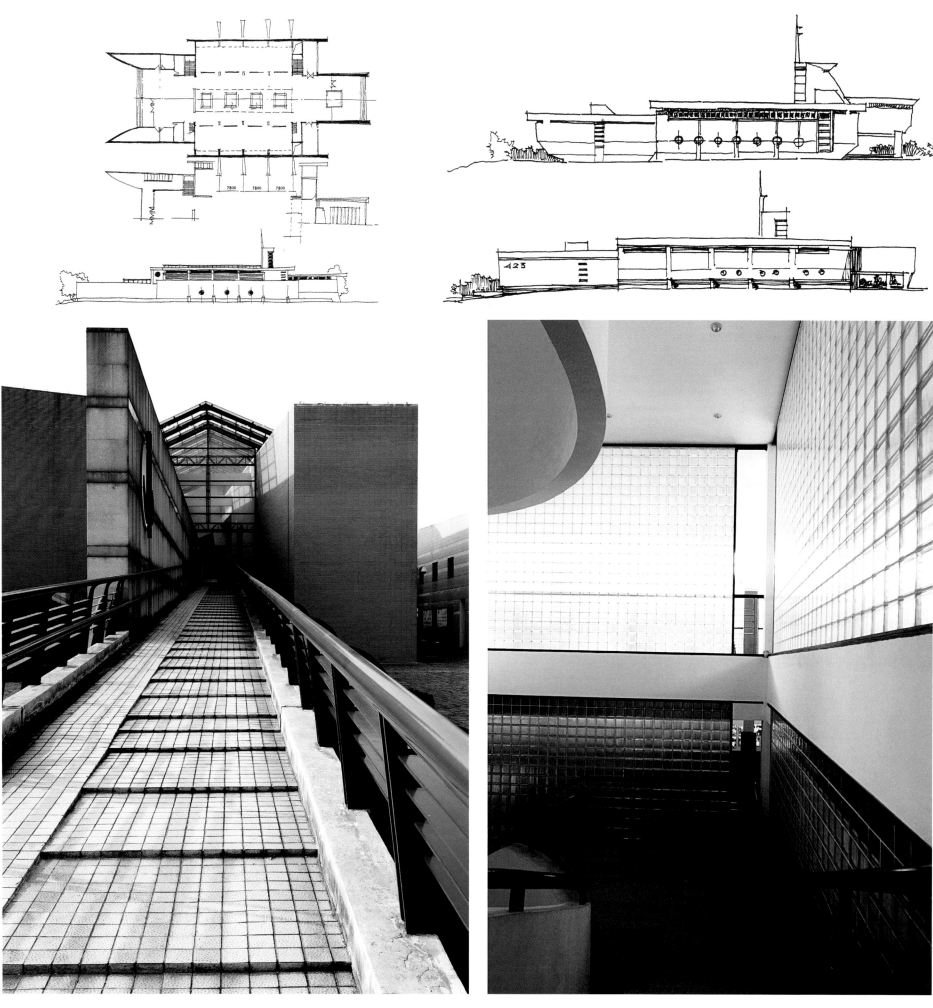

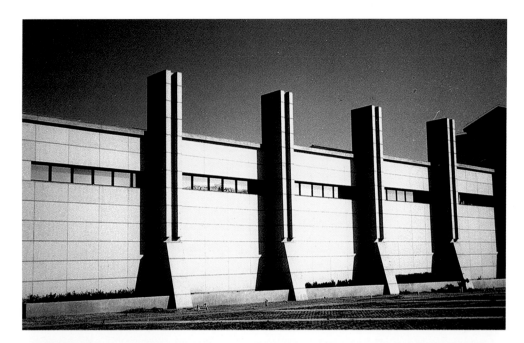

江苏海安苏中七战七捷纪念碑、馆
MEMORIAL OF THE SEVEN VICTORIES IN MID-JIANGSU, HAI'AN JIANGSU PROVINCE

设 计 人：齐康、马啸平、陈泳、张飞
工程地点：江苏省海安县
工程规模：4030平方米
设计时间：1987–1988
建成时间：1988
合作单位：南通市建筑设计研究院
获奖情况：1991年获城乡建设部优秀建筑设计表扬奖

Designer: QI Kang, MA Xiao-ping, CHEN Yong, ZHANG Fei
Location: Haian City, Jiangsu Province
Total Area: 4,030 sq.m.
Design Time: 1987-1988
Completion: 1988
Co-operation: Nantong Institute of Architectural Design
Prize: The Homer Prize of Excellent Architectural Design, Construction Ministry, 1991

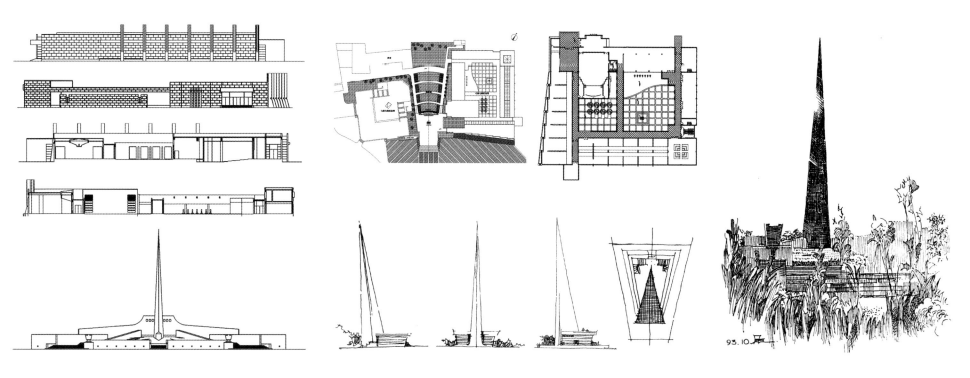

淮海战役陈官庄地区歼灭战烈士陵园
HUAIHAI CAMPAIGN CHENGUANZHUANG ANNIHILATION REGION MARTYRS CEMETERY

设 计 人：齐康、金俊、何柯、张芳、范双丹、
　　　　　徐晨、汪亮
工程地点：河南省永城市
工程规模：40.59公顷
设计时间：2007.8
建成时间：2010
合作单位：东南大学建筑设计研究院

Designer: QI Kang, JIN Jun, HE Ke, ZHANG Fang,
　　　　　FAN Shuang-dan, XU Cheng, WANG Liang
Location: Yongcheng City, Henan Province
Total Area: 40.59 hectares
Design Time: August, 2007
Completion: 2010
Co-operation: Architects & Engineers Co., LTD of
　　　　　Southeast University

设计构思：

　　淮海战役陈官庄地区歼灭战烈士陵园坐落在永城市陈官庄乡，311国道北侧，每年接待来自豫东、皖北、鲁南200平方公里范围的观众约15万人次。是全国重点烈士纪念建筑物保护单位和全国爱国主义教育示范基地，多年来，陵园在褒扬先烈、教育后人方面发挥了重要作用。

　　根据淮海战役陈官庄地区歼灭战烈士陵园改扩建工程领导组的意见，将陵园规划成为以悼念先人及教育后代为核心和主体功能，以风景旅游和生态保护为延伸功能，意在将该景区发展成一个集红色旅游、风景旅游和生态保护为一体的现代化风景区。

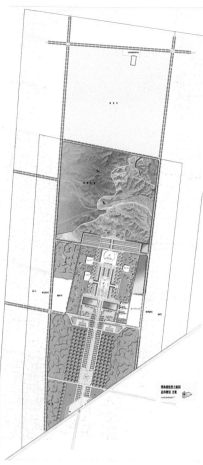

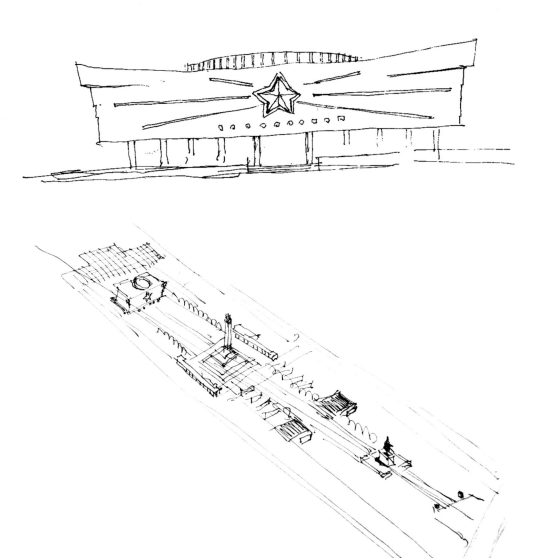

黄山国际大酒店
INTERNATIONAL HOTEL OF HUANGSHAN

设 计 人：齐康、陈宗钦、郑炘	Designer: QI Kang, CHEN Zong-qin, ZHENG Xin
工程地点：安徽省黄山市	Location: Huangshan City, Anhui Province
工程规模：16000 平方米	Total Area: 16,000 sq.m.
设计时间：1990–1992	Design Time: 1990-1992
建成时间：1993	Completion : 1993
合作单位：东南大学建筑设计研究院	Co-operation: Architects & Engineers Co., LTD of Southeast University
获奖情况：1998 年教育部优秀工程设计二等奖 建设部优秀工程设计三等奖	Prize: The Second Prize of Architecture Academic Award,1998 The Third Prize of Architecture Academic Award Construction Ministry

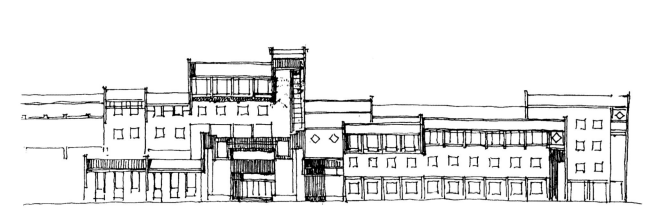

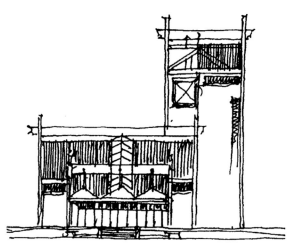

净月潭风景区整体设计
HOLISTIC DESIGN OF JING YUE TAN SCENIC SPOT

设 计 人：齐康、郑炘、王莉、高芸、颜红
工程地点：吉林省长春市
工程规模：186.1 公顷
设计时间：1998–1999
建成时间：2000
合作单位：吉林省冶金设计院

Designer: QI Kang, ZHENG Xin, WANG Li, GAO Yun, YAN Hong
Location: Changchun City, Jilin Province
Total Area: 186.1 hectares
Design Time: 1998-1999
Completion: 2000
Co-operation: Design Instiute of Metallurgy, Jilin Province

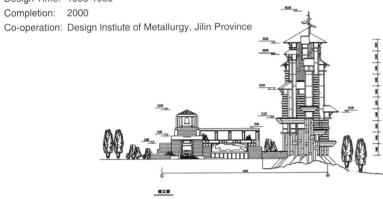

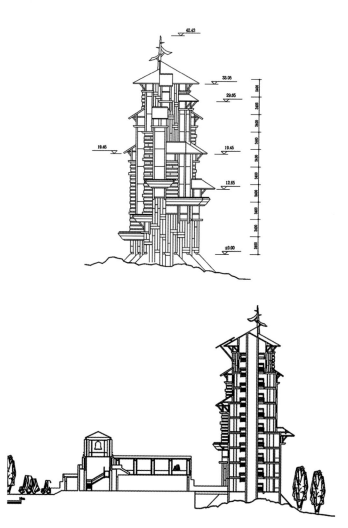

设计构思：

　　景区主入口由售票处和入口标志构成；售票处呈箭头状的屋顶覆盖着弹头状的平面，将斜向的屋脊延伸，并与斜柱交会；入口标志的形象源自新月，造型简洁，线条流畅。两者既分又合，形态上遥相呼应，成为有机整体。

　　观景塔位于原大钟亭边的山顶区域，造型受松柏树形象的启发：塔身挺拔并局部收进，给人以主干之感。坡顶及挑台设于不同的层高，并相互有些错位，辅以斜撑，有如层层树冠一般；塔尖也有树干顶部之感。整体设计具有很强的原创性。

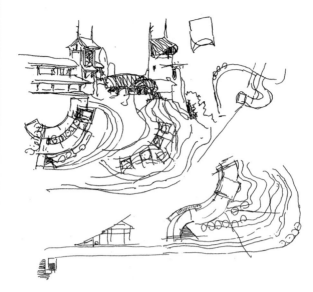

福建长乐海螺塔
CONCH TOWER, CHANGLE, FUJIAN

设 计 人：齐康、郑炘、张宏
工程地点：福建省长乐市
工程规模：200 平方米
设计时间：1986–1987
建成时间：1987
合作单位：南京市建筑设计研究院
获奖情况：1992 年"建筑师杯"全国中小型
　　　　　建筑优秀设计表扬奖

Designer: QI Kang, ZHENG Xin, ZHANG Hong
Location: Changle City, Fujian Province
Total Area: 200 sq.m.
Design Time: 1986-1987
Completion: 1987
Co-operation: Nanjing Institute of Architectural Design
Prize: National Prize of "Architect Cup" for Medium & Small Architectural Works, 1992

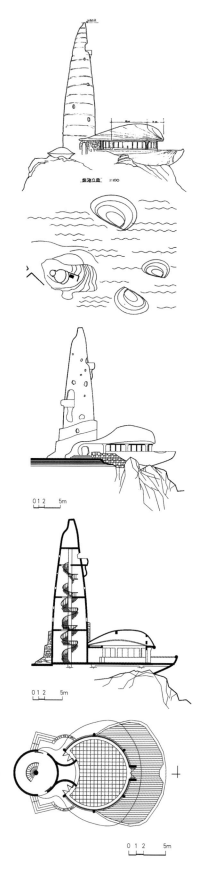

浙江天台山济公佛院
JIGONG TEMPLE OF TIANTAI MOUNTAIN, ZHEJIANG PROVINCE

设 计 人：	齐康、陈宗钦、陈公余
工程地点：	浙江省天台县
工程规模：	500 平方米
设计时间：	1986
建成时间：	1987
合作单位：	天台县建筑设计室
获奖情况：	1992 年"建筑师杯"全国中小型建筑优秀设计优秀奖 浙江省优秀景点一等奖

Designer:	QI Kang, CHEN Zong-qin, CHEN Gong-yu
Location:	Tiantai County, Zhejiang Province
Total Area:	500 sq.m.
Design Time:	1986
Completion:	1987
Co-operation:	Architectural Institute of Tiantai County
Prize:	National Merit Prize of "Architect Cup" for Medium & Small Architecture Works, 1992 The First Prize of Zhejiang Province Excellent Tourist Spots

福建惠安海门亭
HUIAN HAIMEN PAVILION, FUJIAN PROVINCE

设 计 人：齐康
工程地点：福建省惠安县
设计时间：1986
建成时间：1987

Designer: QI Kang
Location: Huian City, Fujian Province
Design Time: 1986
Completion: 1987

江阴望江楼
BELVEDERE IN JIANGYIN

设 计 人：齐康、齐昉
工程地点：江苏省江阴市
工程规模：1200平方米
设计时间：1991–1992
建成时间：1993
合作单位：东南大学建筑设计研究院

Designer: QI Kang, QI Fang
Location: Jiangyin City, Jiangsu Province
Total Area: 1,200 sq.m.
Design Time: 1991-1992
Completion : 1993
Co-operation: Architects & Engineers Co., LTD of Southeast University

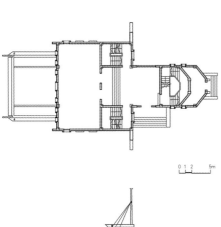

西藏和平解放纪念碑
THE MONUMENT FOR THE PEACEFUL LIBERATION OF TIBET

设 计 人：齐康、张宏、叶菁、孙磊磊、温洋
工程地点：西藏自治区拉萨市
工程规模：3.6公顷（广场用地）
设计时间：2001.3–2001.8
建成时间：2002
合作单位：东南大学建筑设计研究院
获奖情况：教育部2003年度优秀勘察设计一等奖

Designer: QI Kang, ZHANG Hong, YE Jing, SUN Lei-lei, WEN Yang
Location: Lasa City, Tibet
Total Area: 3.6 hectares(Plaza)
Design Time: March, 2001-August, 2001
Completion : 2002
Co-operation: Architects & Engineers Co., LTD of Southeast University
Prize: The First Price for Excellent Survey and Design, Education Ministry, 2003

设计构思：

西藏和平解放纪念碑位于拉萨市布达拉宫南广场南端，西为文化宫，东为现有保留水面，南为自治区政府，北临幸福路。广场规划总用地面积约3.6公顷。碑体造型是从世界最高峰珠穆朗玛峰的形象上获得灵感，借用其高耸入云的气势，与天地同在的永恒性，以建筑化、抽象化的语汇来进行创作。纪念碑主体以灰白色为主色调，不过多装饰，挺拔、简洁、浑然一体，从气势上体现了西藏和平解放、农奴翻身做主人中所蕴含的伟大的具有世界性的精神，具有极强的震撼力与艺术感染力，意义深远。

纪念碑底部基座高3米，采用草坡形式，结合跌落式片墙，使纪念碑犹如从大地中生长出来，庄严、神圣。前方设计有两组大理石群雕，以西藏农奴翻身得解放、解放军筑路为主题。在纪念碑的细部处理上，碑身局部嵌入不规则的金色、红色镜面玻璃细带，与布达拉宫的色彩相呼应。入口门洞上方和基座处饰以带有地方风格的装饰和色彩，使整体造型更为成熟与完善。

纪念碑内部入口处上方以向上层层收缩的方式一直通到顶部，设有天窗，阳光透过各色玻璃，从上方及四面墙体上的条带窗中投射进来，给人以无限遐想，形成极富表现力和纪念性的空间艺术氛围。

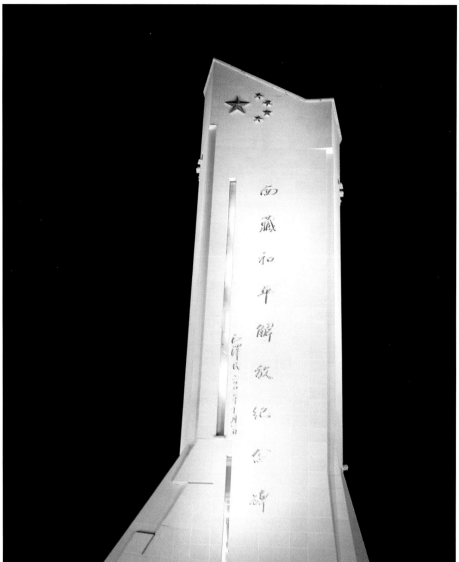

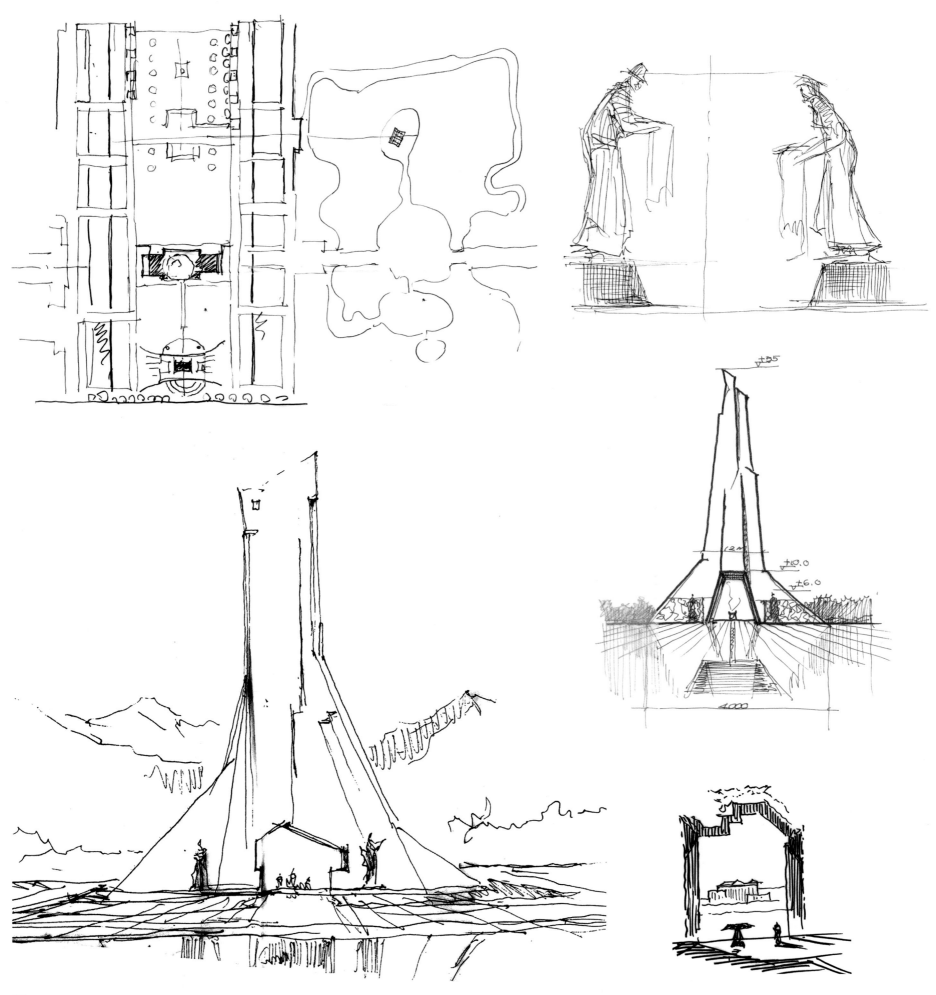

福建省博物馆
MUSEUM OF FUJIAN PROVINCE

设 计 人：齐康、林卫宁、杨志疆、邓浩
工程地点：福建省福州市
工程规模：35000 平方米
设计时间：1997.9–1998.5
建成时间：2002.10
合作单位：福建省建筑设计研究院

Designer: QI Kang, LIN Wei-ning,
 YANG Zhi-jiang, DENG Hao
Location: Fuzhou City, Fujian Province
Total Area: 35,000 sq.m.
Design Time: September, 1997-May, 1998
Completion: October, 2002
Co-operation: Fujian Institute of Architectural Design

设计构思：

福建省博物馆是集历史博物馆、自然博物馆、闽台交流中心、积萃园艺术馆和考古研究所等为一体的综合性博物馆。新馆位于西湖公园内，三面环水，总用地面积约 89 亩。

在总平面的设计中，强调了城市尺度与环境尺度的相结合，使其在整合这一区域的同时，成为一种"包容性"的中心，既融于环境，又创造环境，并最终成为西湖公园中的一座"博物花园"。

在平面功能的设计中，博物馆被划分为主馆区、综合馆区、自然馆区和检测中心区四个主要的部分。主入口设计在二层，这样从地面广场到主入口便形成了宽阔的台阶平台，结合浮雕墙塑造出浓郁的文化氛围，形成了空间的第一层次；进入室内后以序言厅为中轴，用玻璃天棚将阳光引入室内，形成空间的第二层次；经过序言厅进入中央大厅，各陈列厅围绕大厅布置，近20m的中庭成为整个建筑空间序列的高潮所在。

在形态景观的设计中，没有简单地套用原有的传统形式，而是抽象出福建民居飞檐大量并置而产生的层层相叠的形象特征，在整个造型的水平方向，依据各个体块的层层相契的关系加以运用，得到了地方形式的固有的韵律特点。而在主入口、单元体等处又重点地研究了地方曲线的现代性处理，结合近40米高的图腾柱的隐喻，表达出一种浑厚的、具有地方特质的文化内涵。

哈尔滨金上京历史博物馆
HISTORICAL MUSEUM OF JING CAPITAL, HARBIN

设 计 人：齐康、郑炘、刘红杰
工程地点：黑龙江省哈尔滨市
工程规模：4000平方米
设计时间：1997
建成时间：1998
合作单位：哈尔滨建筑大学建筑设计研究院

Designer:	QI Kang, ZHENG Xin, LIU Hong-jie
Location:	Harbin City, Heilongjiang Province
Total Area:	4,000 sq.m.
Design Time:	1997
Completion:	1998
Co-operation:	Architectural Design and Research Institute of Harbin Architectural College

设计构思：

博物馆结合周围环境采用四合院布局形式，东面入口大门以其东向的朝向与历史故都会宁府取得空间上的对应，西北以打开四合院的缺口与历史遗留的阿骨打墓冢取得空间上的斜向关系。

带有某种"武士"寓意的方形中庭，犹如刀枪行列的入口架，以及象征"烽火台"的西北出口均使博物馆具有强烈的标志性形象，充分表达了地区的、历史的文化。

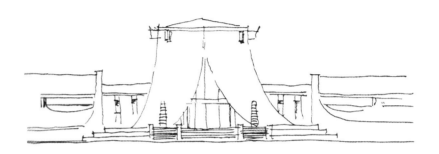
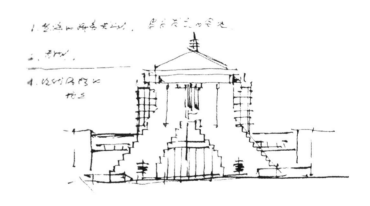

冰心文学馆
BINGXIN MEMORIAL

设 计 人：齐康、林卫宁、王建国
工程地点：福建省长乐市
工程规模：4413 平方米
设计时间：1996
建成时间：1997.8
合作单位：福建省建筑设计研究院
获奖情况：1998 年福建省优秀工程设计一等奖
　　　　　1999 年福建省双十佳建筑奖

Designer: QI Kang, Lin Wei-ning, WANG Jian-guo
Location: Changle City, Fujian Province
Total Area: 4,413 sq.m.
Design Time: 1996
Completion: August, 1997
Co-operation: Fujian Institute of Architectural Design
Prize: The First Prize of Fujian Provincial Architecture
             Academic Award, 1998
             Fujian Top20 Architectural Award, 1999

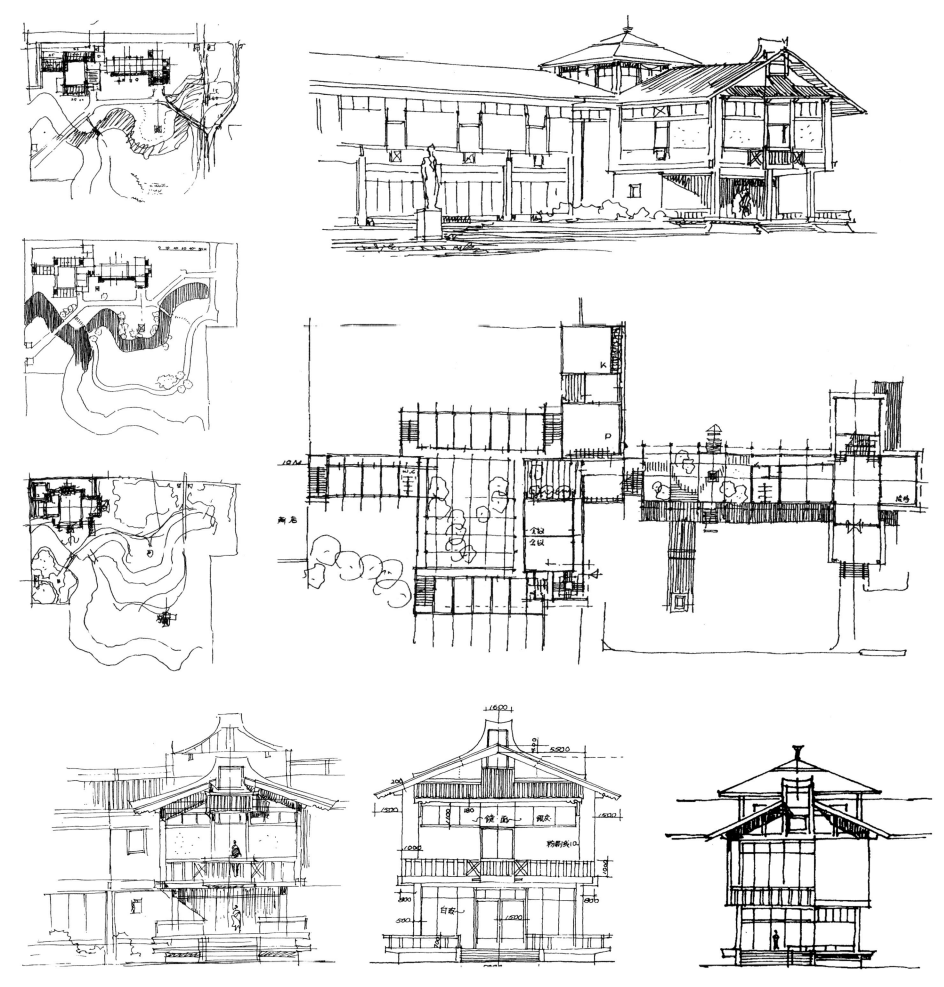

华罗庚纪念馆
HUA LUOGENG MEMORIAL

设 计 人：齐康、华晓宁、叶菁、张四维
工程地点：江苏省常州市
工程规模：2622平方米
设计时间：2004.9
建成时间：2008
合作单位：东南大学建筑设计研究院

Designer: QI Kang, HUA Xiao-ning, YE Jing, ZHANG Si-wei
Location: Changzhou City, Jiangsu Province
Total Area: 2,622 sq.m.
Design Time: September, 2004
Completion: 2008
Co-operation: Architects & Engineers Co., LTD of Southeast University

设计构思：

　　华罗庚纪念馆的创作在充分理解周边自然环境和地域文化的基础上，以数学几何的平面构成、简约明快的现代造型、宁静质朴的空间氛围，形成独特的建筑气质，以内在的象征性与纪念性，凝结成一座人文精神的展示场，充分展现出华罗庚先生的事业成就与人格精神。

　　建筑在基本位置的选定、平面结构的走向以及空间体量的伸展上，与环境中既有的建筑、河流、水面、道路形成极为密切的关系，几乎是完美地"嵌入"环境结构之中。

　　圆形的中庭，四角锥形的玻璃天窗，呈水平、垂直45°角插入的外墙，均呈现出几何的构成、数学的隐喻。几何化的建筑形体、高低错落的建筑体量、白色的花岗石墙体、细长的横向水平条窗以及强烈的虚实对比，在绿树和水面的环境下，整体造型简洁而现代、平淡而朴实。

苏州丝绸博物馆
MUSEUM OF SILK IN SUZHOU

设 计 人：齐康、宋淑敏、马啸平、童明	Designer: QI Kang, SONG Shu-min, MA Xiao-ping, TONG Ming
工程地点：江苏省苏州市	Location: Suzhou City, Jiangsu Province
工程规模：8000 平方米	Total Area: 8000 sq.m.
设计时间：1989–1990	Design Time: 1989-1990
建成时间：1990	Completion: 1990
合作单位：苏州市建筑设计研究院	Co-operation: Suzhou Institute of Architectural Design
获奖情况：1993年建设部优秀设计三等奖	Prize: The Third Prize of Architecture Academic Award, Constrution Ministry,1993
1993年江苏省优秀设计一等奖	The First Prize of Jiangsu Provincial Architecture Academic Award, 1993

苏州丝绸博物馆改扩建
MUSEUM OF SILK EXPANSION DESIGN, SUZHOU

设 计 人：齐康、杨志疆、叶菁、邵继中	Designer: QI Kang, Yang Zhi-jiang, Ye Jing, Shao Ji-zhong
工程地点：江苏省苏州市	Location: Suzhou City, Jiangsu Province
工程规模：2539 平方米	Total Area: 2539 sq.m.
设计时间：2012–2013	Design Time: 2012-2013
建成时间：2014	Completion: 2014
合作单位：东南大学建筑设计研究院有限公司	Co-operate: Architects & Engineers Co., LTD of Southeast University

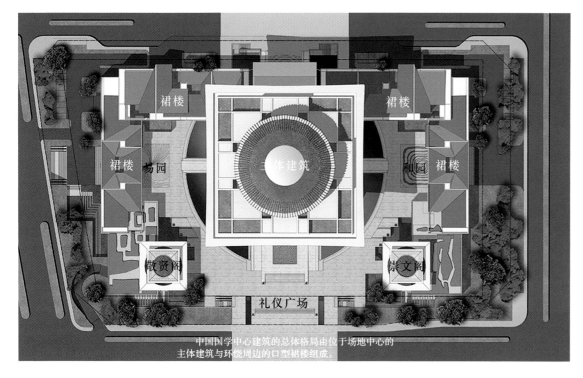

中国国学中心建筑的总体格局由位于场地中心的主体建筑与环绕周边的口型裙楼组成。

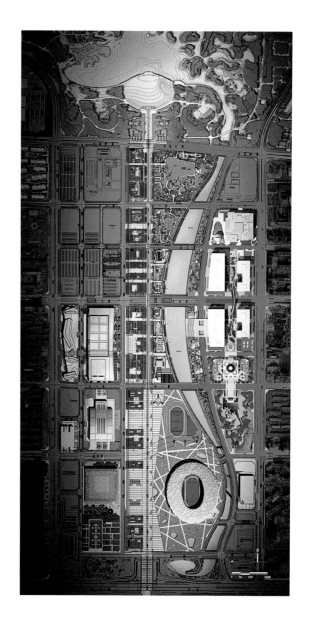

中国国学中心
CHINA NATIONAL SINOLOGY CENTER

设 计 人：齐康、王建国、张彤
工程地点：北京市
工程规模：82500平方米
设计时间：2012—2014
建成时间：2017

Designer: QI Kang, WANG Jian-guo, ZHANG Tong
Location: Bbeijing City
Total Area: 82500 sq.m.
Design Time: 2013-2014
Completion: 2017

设计构思：

总体布局采用中国传统建筑最为壮丽的宫阙形制，以裙楼环抱主体建筑，以东西两阙衬托中央耸立之势。在体现国家宏硕壮美之文化形象的同时，形成一系列亲民开放的外部空间场所，与文化综合区的公共空间系统衔接融合。

建筑空间型制秉承东方哲学独有的时空图式，源承以明堂辟雍为代表的中国最高规格精神性建筑的方圆叠套几何形态，凸显中心感、对称性和宇宙象征意义。空间设计以国子监辟雍和曲阜孔庙杏坛为原型，彰显"圜桥教泽、传流四方"的意义。

建筑造型以中国古典建筑最具特征的经典线型，即由斗拱承托从屋檐到柱身的静力学传力曲线为造型母题。在竖向的韵律中，自下而上呈现出由实到虚、由简而繁的渐次变化。以纯美之形态、升腾之韵律展现中华国学中正淳和的精神气质，彰显民族文化大繁荣大发展的气势。

113

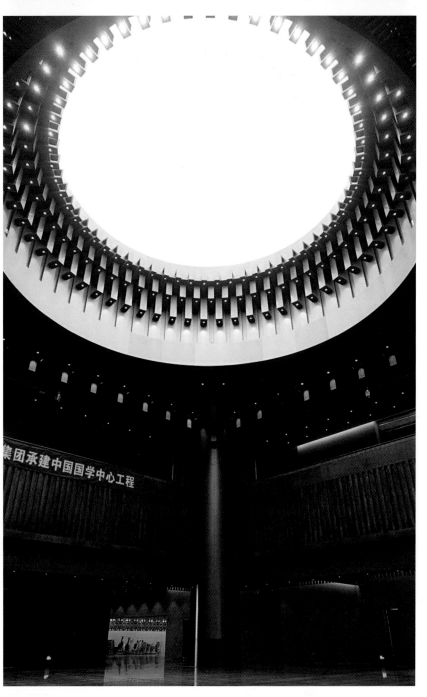

南京 1912 民国文化街区
1912 CHINA CULTURAL DISTRICT, NANJING

设 计 人：杨志疆、齐康、寿刚等
工程地点：江苏省南京市
工程规模：20000 平方米
设计时间：2002
建成时间：2004
合作单位：东南大学建筑设计研究院
获奖情况：2006 年建设部全国优秀工程设计银奖

Designer: YANG Zhi-jiang, QI Kang, SHOU Gang, etc
Location: Nanjing City, Jiangsu Province
Total Area: 20,000 sq.m.
Design Time: 2002
Completion: 2004
Co-operation: Architects & Engineers Co., LTD of Southeast University
Prize: The Silver Prize of Architecture Academic Award, Construction Ministry, 2006

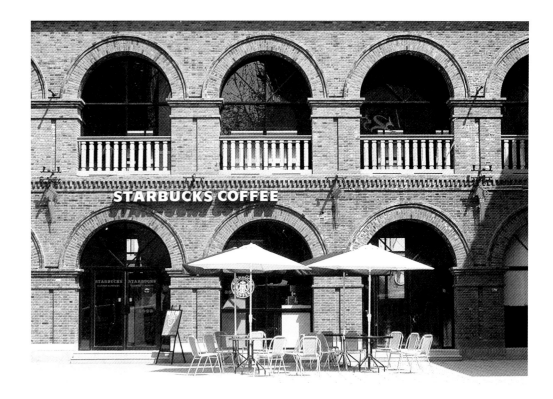

设计构思：
　　"南京1912"位于"总统府"西北侧，总占地约3.5公顷。在这一用地区域内包涵了丰厚浓郁的文化氛围，具有极高的人文价值、文化价值和历史价值。街区的功能定位是"总统府"近代史博物馆的配套服务设施，希望借助于这一地段在城市中所特有的文化属性及历史渊源来构筑集餐饮、娱乐、观光于一体的文化型商业街区。
　　"总统府"旧址内现存的建筑混合交织了从清末到民国的各个时期的多种类型的建筑样式。为此，我们通过提炼形态设计上的一些基本原型（如屋顶的形式、"拱券"和青砖的质感），将过去、现在和未来交织在一起，令人遐想和深思。

中国鞋文化博物馆
THE CHINESE SHOES CULTURE MUSEUM, WENZHOU

设 计 人：齐康、叶菁、齐昉	Designer: QI Kang, YE Jing, QI Fang
工程地点：浙江省温州市	Location: Wenzhou City, Zhejiang Province
工程规模：1230 平方米	Total Area: 1,230 sq.m.
设计时间：2000.6–2000.10	Design Time: June, 2000-October, 2000
建成时间：2001.11	Completion: November, 2001
合作单位：东南大学建筑设计研究院	Co-operation: Architects & Engineers Co., LTD of Southeast University
获奖情况：2003 年度南京市优秀工程设计二等奖	Prize: The Second Prize of Nanjing Outstanding Engineering Design, 2003

设计构思：

中国鞋文化博物馆位于温州市区西部双屿工业区卧旗山山顶，该山被瓯江三面环绕。

建筑在整体构思中，将建筑高度控制在 12 米，并面向江面以弧线形轨迹倾斜至地面标高，在体现"鞋"形的同时，与山体紧密契合，与瓯江谦逊对话。

建筑通过曲线的运用、历史的表达、语汇的抽象等多种手法，形成独特的造型，带给观赏者以视觉上的快感。设计以"鞋"为主题，采用抽象与具象相结合的手法将多重的象征意义糅合到建筑

语言当中。平面布局、立体造型以"鞋"特有的形体为意象，加以抽象、延伸——不对称、曲线、倾斜、统一色彩等。高18米的不锈钢雕塑将人体美与鞋的柔性美相结合，以含混、模糊的理念来体现未来鞋文化的发展和人类对美的憧憬。入口雨篷、博物馆标志、斜拉索等细部处理，更是遵循同样的设计原则，使整体与局部风格相统一。

钢结构、铝屋面与玻璃天窗的运用使建筑同时体现了时代的特色。特别是在室内设计中，更突出了这一点：在简洁统一的白色墙面与灰色地面衬托下，玻璃与钢以其特有的现代感成为空间氛围的创造者，声音、光线的介入，电子设备的运用均赋予博物馆很强的时代气息。

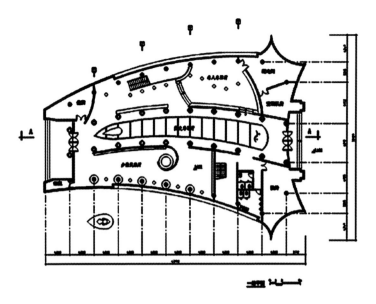

大连贝壳博物馆
DALIAN CONCH MUSEUM, LIAONING PROVINCE

设 计 人：齐康、胡文荟、刘媛	Designer: QI Kang, HU Wen-hui, LIU Yuan
工程地点：辽宁省大连市	Location: Dalian City, Liaoning Province
工程规模：10000 平方米	Total Area: 10,000 sq.m.
设计时间：2006.2–2006.8	Design Time: February, 2006-August, 2006
建成时间：在建	Completion: Under Construction
合作单位：大连理工大学土木建筑设计研究院	Co-operation: The Design Institute of Civil Engineering & Architecture of DUT

设计构思：

　　大连市位于中国东北的最南端，东临黄海，西邻渤海，南与山东半岛隔海相望，北与东北大陆相连，是一座三面环海的半岛城市。大连全地区海岸线长 1906 公里，南部海滨风景名胜区是大连绚丽海滨的精华。

　　大连市贝壳博物馆馆址位于大连市西南部的马栏河畔，毗邻大连市著名的星海广场，总用地面积约 10000 平方米。计划建成后将引进世界各地大量珍奇贝类品种，成为集收藏、研究、展览为一体的大型贝壳专业博物馆，并成为大连市的地区文化标志之一。

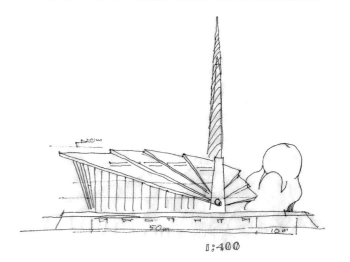

  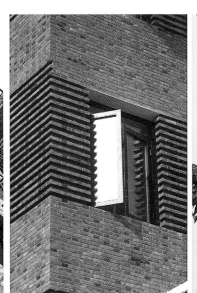  

晨光 1865 东部入口区
EAST ENTRANCE ZONE OF CHENGUANG 1865, NANJING

设 计 人：王彦辉、齐康、杨程、朱亚楠
工程地点：江苏省南京市
工程规模：2205 平方米
设计时间：2007
建成时间：2009

Designer: WANG Yan-hui, QI Kang, YANG Cheng, ZHU Ya-nan
Location: Nanjing City, Jiangsu Province
Total Area: 2,205 sq.m.
Design Time: 2007
Completion: 2009

南京中信大楼
ZHONGXIN TOWER, NANJING

设 计 人：齐康、邹式汀	Designer: QI Kang, ZOU Shi-ting
工程地点：江苏省南京市	Location: Nanjing City, Jiangsu Province
工程规模：4000平方米	Total Area: 4,000 sq.m.
设计时间：1993–1998	Design Time: 1993-1998
建成时间：2000	Completion: 2000
合作单位：南京市建筑设计研究院	Co-operation: Nanjing Institute of Architectural Design
获奖情况：2001年南京市优秀工程设计一等奖	Prize: The First Prize of Nanjing Architecture Academic Award, 2001
2001年江苏省优秀工程设计三等奖	The First Prize of Jiangsu Provincial Architecture Academic Award, 2001

南京达舜国际广场
DASHUN INTERNATIONAL SQUARE, NANJING

设 计 人：齐康、邹式汀、张四维等
工程地点：江苏省南京市
工程规模：26472 平方米
设计时间：2004.3
建成时间：2007
合作单位：南京市建筑设计研究院

Designer:　　QI Kang, ZOU Shi-ting,
　　　　　　　ZHANG Si-wei, etc
Location:　　Nanjing City, Jiangsu Province
Total Area:　　26,472 sq.m.
Design Time: March, 2004
Completion:　2007
Co-operation: Nanjing Institute of Architectural Design

中国人民银行南京分行
NANJING BRANCH OF PEOPLE'S BANK OF CHINA

设 计 人：齐康、段进、蒋桂泉、陈泳、齐昉
工程地点：江苏省南京市
工程规模：58700 平方米
设计时间：1992–1995
建成时间：1997
合作单位：东南大学建筑设计研究院
获奖情况：2000 年教育部优秀设计表扬奖

Designer: QI Kang, DUAN Jin, JIANG Gui-quan, CHEN Yong, QI Fang
Location: Nanjing City, Jiangsu Province
Total Area: 58,700 sq.m.
Design Time: 1992-1995
Completion: 1997
Co-operation: Architects & Engineers Co., LTD of Southeast University
Prize: Honer Award Excellent Design, Education Ministry, 2000

南京鼓楼邮政大楼
GULOU POST MANSION, NANJING

设 计 人：	齐康、刘照泓、张俊仪、仲德崑、王建国
工程地点：	江苏省南京市
工程规模：	40000 平方米
设计时间：	1992–1994
建成时间：	1997.5
合作单位：	南京市建筑设计研究院
获奖情况：	1998年江苏省优秀工程设计一等奖
	1998年建设部优秀工程设计三等奖

Designer:	QI Kang, LIU Zhao-hong, ZHANG Jun-yi, ZHONG De-kun, WANG Jian-guo
Location:	Nanjing City, Jiangsu Province
Total Area:	40,000 sq.m.
Design Time:	1992-1994
Completion:	May, 1997
Co-operation:	Nanjing Institute of Architectural Design
Prize:	The First Prize of Jiangsu Provincial Architecture Academic Award, 1998
	The Third Prize of Architecture Academic Award, Construction Ministry, 1998

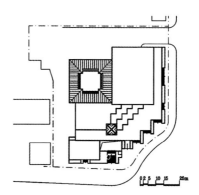
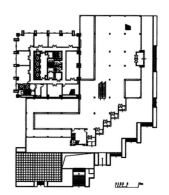
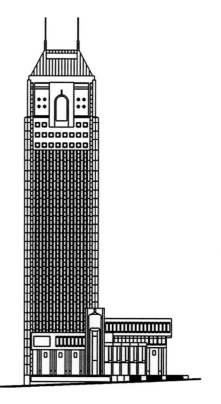

设计构思：

邮政大楼是一座现代化建筑，因此形象特征在传达传统的意蕴的同时，也要反映现代的科技。造型处理汲取现代建筑处理的一些手法，从强调垂直划分（这是现代高层的通常手法之一）过渡到拱门。从方案设计推敲过程中，不难想到北京前门楼上的"方窗"洞。拱门是最宜代表钟楼的象征，是北京钟楼的形的重影，正方形的塔楼，自然引起顶部四面向上坡。坡度的陡和缓是研究形象所必须考虑的，同时也要考虑斜的对景，最终形成了现有的形象，最高点构思用中国传统塔的"刹"来作为点缀。墙面采用仿古面砖，效果很好。裙房则运用现代手法转化为接近街道和人活动可以觉察的尺度。最后在顶部门拱的玻璃采用了红色。

形象特征考虑到地段的文化特点，特别是现有的南京主要历史标志性建筑之一的鼓楼。中国古代城市规则有"左鼓，右钟"的形制。我们的构思设想将主楼的层顶设想为"钟楼"，以"钟"来表达其内涵。具体设计时我们剖析了威尼斯广场的Capnile钟楼、伦敦West Minster Abbey的钟楼、北京钟鼓楼，以及有关中国寺庙钟鼓楼的形象特征。

从总体上我们研究了环境的尺度、视觉和视距，分析不同视觉的观赏，在自然环境意识、人造空间环境意识、历史文化意识和现代持续发展的意识基础上创造一种文脉的延续。

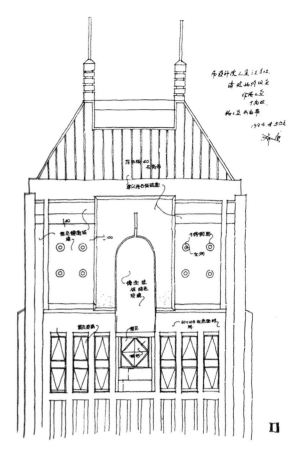

南京华侨大厦
OVERSEA CHINESE HOTEL, NANJING

设 计 人：齐康、王澍
工程地点：江苏省南京市
工程规模：12000 平方米
设计时间：1985–1987
建成时间：1987
合作单位：南通市建筑设计研究院

Designer: QI Kang, WANG Shu
Location: Nanjing City, Jiangsu Province
Total Area: 12,000 sq.m.
Design Time: 1985-1987
Completion: 1987
Co-operation: Nantong Institute of Architectural Design

南京怡华假日酒店
YIHUA HOTEL, NANJING

设 计 人：齐康、赖聚奎、王建国
工程地点：江苏省南京市
工程规模：18000 平方米
设计时间：1995
建成时间：1997

Designer: QI Kang, LAI Ju-kui, WANG Jian-guo
Location: Nanjing City, Jiangsu Province
Total Area: 18,000 sq.m.
Design Time: 1995
Completion: 1997

江苏省国税大厦
NATIONAL TAX BUILDING, JIANGSU PROVINCE

设 计 人：齐康、齐昉、王建国、华峰	Designer: QI Kang, QI Fang, WANG Jian-guo, HUA Feng
工程地点：江苏省南京市	Location: Nanjing City, Jiangsu Province
工程规模：27000 平方米	Total Area: 27,000 sp.m.
设计时间：1995	Design Time: 1995
建成时间：1999	Completion: 1999
合作单位：东南大学建筑设计研究院	Co-operation: Architects & Engineers Co., LTD of Southeast University
获奖情况：2000 年江苏省优秀工程设计二等奖	Prize: The Second Prize of Jiangsu Provincial Architecture Academic Award, 2000

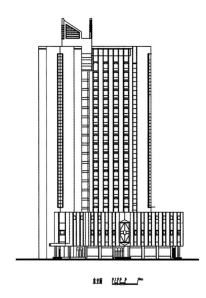

中国人民银行济南分行营业楼
JINAN BRANCH OF THE PEOPLE'S BANK OF CHINA

设 计 人：齐康、郑炘、齐昉、杨志疆、
　　　　　于雷、史晓州
工程地点：山东省济南市
工程规模：50300 平方米
设计时间：1997–1999
建成时间：2001.11
合作单位：东南大学建筑设计研究院
获奖情况：2003 年江苏省优秀工程设计二等奖

Designer: QI Kang, ZHENG Xin, QI Fang,
　　　　　YANG Zhi-jiang, YU Lei, SHI Xiao-zhou
Location: Jinan City, Shandong Province
Total Area: 50,300 sq.m.
Design Time: 1997-1999
Completion: November, 2001
Co-operation: Architects & Engineers Co., LTD of Southeast University
Prize: The Second Prize of Outstanding Engineering Design Jiangsu Province, 2003

设计构思：

　　该建筑位于济南市中心繁华的金融街上，基地为一不规则的长方形，其中经七路与纬四路之间形成 12°的夹角。由于基地周围现存的金融办公建筑一味追求高度，造成沿街建筑界面的断裂与街区感的削弱。本方案以板式高层与裙房围合成"U"形体量，这样一方面可以与相邻街区建筑体量相呼应，通过积极地介入使失序的街区空间重新有序化；另一方面由于在群体内部形成一个庭院空间，底层公共营业场所显得更加人性化。

　　板式高层的形式反映了内部空间的特点，大面积的铝框玻璃幕墙暗示了办公空间，它与凸出于立面、集中了各类辅助性功能的服务体实墙面形成了强烈的对比。整个建筑立面的纵向采用了古典建筑的段式处理，顶部是各种机房，中部为办公标准层，底部是不同用途的公共空间，视觉上的稳定感有助于建立公众对银行的信心。在剖面设计中，主楼每三层设一处共享空间，增加私密办公场所中的公共交往活动。

深圳国人大厦
GUOREN BUILDING OF SHENZHEN

设 计 人：齐康、郑炘、桂鹏、朱巍
工程地点：广东省深圳市
工程规模：73000 平方米
设计时间：2006
建成时间：2008
合作单位：深圳市建筑设计院

Designer:	QI Kang, ZHENG Xin, GUI Peng, ZHU Wei
Location:	Shenzhen City, Guangdong Province
Total Area:	73,000 sq.m.
Design Time:	2006
Completion:	2008
Co-operation:	Architectural and Design Institute of Shenzhen

南京古生物博物馆
MUSEUM OF PALEOBIOLOGY, NANJING

设 计 人：齐康、齐昉、邓浩
工程地点：江苏省南京市
工程规模：8553平方米
设计时间：1998–2001
建成时间：2004
合作单位：东南大学建筑设计研究院
获奖情况：2005年教育部优秀建筑设计二等奖
　　　　　2005年建设部优秀勘察设计三等奖

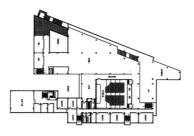

Designer:　　QI Kang, QI Fang, DENG Hao
Location:　　Nanjing City, Jiangsu Province
Total Area:　 8,553 sq.m.
Design Time: 1998-2001
Completion:　2004
Co-operation: Architects & Engineers Co., LTD
　　　　　　　of Southeast University
Prize:　　　　The Second Prize of Architecture Academic
　　　　　　　Award, Education Ministry, 2005
　　　　　　　The Third Prize of Architecture Academic
　　　　　　　Award, Construction Ministry, 2005

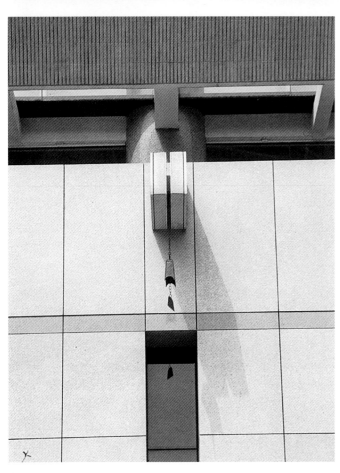

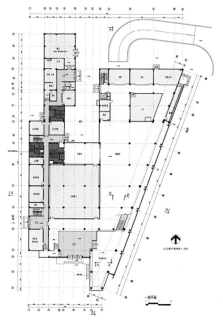
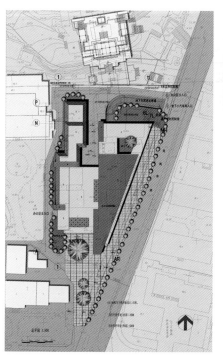

徐州兵马俑博物馆
TERRACOTTA WARRIORS AND HORSES MUSEUM, XUZHOU

设 计 人：齐康、张宏、周曦
工程地点：江苏省徐州市
工程规模：5000 平方米
设计时间：2004
建成时间：2005
合作单位：徐州市第二建筑设计院

Designer: QI Kang, ZHANG Hong, ZHOU Xi
Location: Xuzhou City, Jiangsu Province
Total Area: 5,000 sq.m.
Design Time: 2004
Completion: 2005
Co-operation: The Second Institute of Architectural Design, Xuzhou City

长白山满族文化博物馆
CHANGBAI MOUTAIN MANCHU CULTURE MUSEUM

设 计 人：齐康、刘媛、颜红	Designer: QI Kang, LIU Yuan, YAN Hong
工程地点：吉林省白山市	Location: Baishan City, Jilin Province
工程规模：4770平方米	Total Area: 4,770 sq.m.
设计时间：2006.5	Design Time: May, 2006
建成时间：2008	Completion: 2008
合作单位：吉林省黑色冶金设计院	Co-operation: Ferrous Metallurgy and Design Institute of Jilin Province

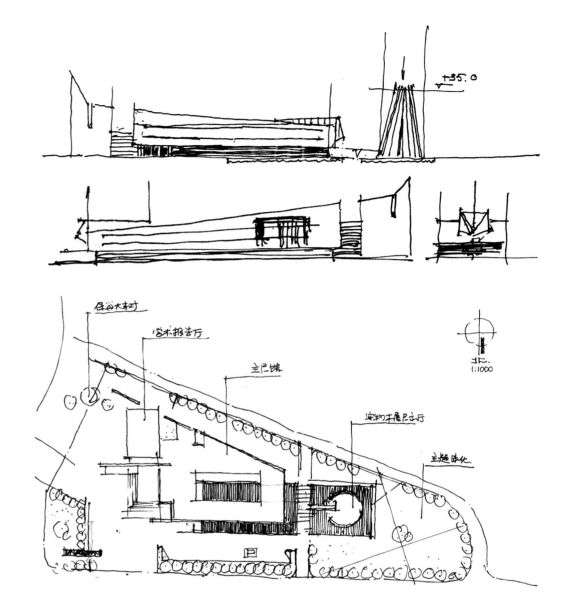

东莞东江纵队纪念馆
THE MEMORIAL HALL OF DONGJIANG COLUMN, DONGGUAN

设 计 人：齐康、郑炘、张四维、谢暾	Designer: QI Kang, ZHENG Xin, ZHANG Si-wei, XIE Tun
工程地点：广东省东莞市	Location: Dongguan City, Guangdong Province
工程规模：3630 平方米	Total Area: 3,630 sq.m.
设计时间：2003.10	Design Time: October, 2003
建成时间：2005.9	Completion: September, 2005

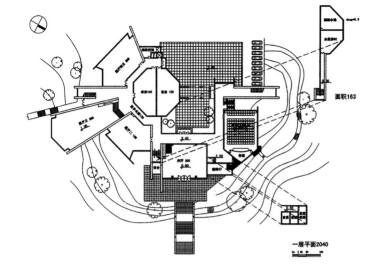

东北沦陷史陈列馆
THE MUSEUM FOR THE HISTORY OF OCCUPATION IN NORTHEAST OF CHINA

设 计 人：郑炘、齐康、谢暾、孙承磊 　　　　　钱玉斋、朱巍	Designer:　ZHENG Xin, QI Kang, 　　　　　XIE Tun, SUN Cheng-lei, 　　　　　QIAN Yu-zhai, ZHU Wei
工程地点：吉林省长春市	Location:　Changchun City, 　　　　　Jilin Province
工程规模：5470 平方米	Total Area:　5,470 sq.m.
设计时间：2004.4	Design Time:　April, 2004
建成时间：2007	Completion:　2007

吉林省集安市高句丽博物馆
KOGURYO HERITAGE MUSEUM IN JIAN, JILIN PROVINCE

设 计 人：齐康、杨程、史庆超、颜红	Designer: QI Kang, Yang Cheng, Shi Qing-chao, YAN Hong
工程地点：吉林省集安市	Location: Jian City, Jilin Province
工程规模：4302平方米	Total Area: 4,302 sq.m.
设计时间：2006.7	Design Time: July, 2006
建成时间：2009.8	Completion: August, 2009
合作单位：长春市城乡规划院	Co-operation: Changchun Urban and Town Planning Institute

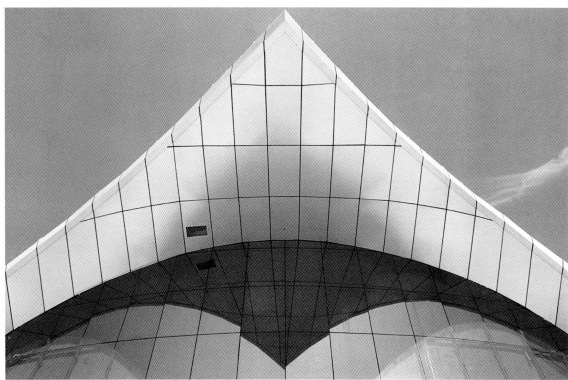

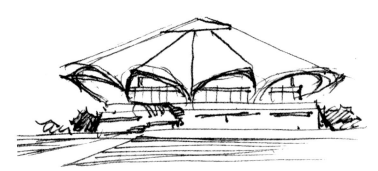

安徽凤阳县小岗村"大包干"纪念馆
DABAOGAN MUSEUM, FENG YANG, ANHUI PROVINCE

设 计 人：齐康、叶菁、钱玉斋、温洋（雕塑）
工程地点：安徽省滁州市
工程规模：2380 平方米
设计时间：2004
建成时间：2006.6
合作单位：东南大学建筑设计研究院

Designer:　　QI Kang, YE Jing, QIAN Yu-zhai,
　　　　　　WEN Yang (Sculpture)
Location:　　Quzhou City, Anhui Province
Total Area:　 2,380 sq.m.
Design Time: 2004
Completion:　June, 2006
Co-operation: Architects & Engineers Co., LTD
　　　　　　of Southeast University

东南大学九龙湖校区图书馆
LIBRARY OF JIULONGHU CAMPUS, SOUTHEAST UNIVERSITY

设 计 人：齐康、齐昉、叶菁、张四维、
　　　　　朱巍等
工程地点：江苏省南京市
工程规模：52900 平方米
设计时间：2004.10
建成时间：2007
合作单位：东南大学建筑设计研究院

Designer: QI Kang, QI Fang, YE Jing,
　　　　　ZHANG Si-wei, ZHU Wei, etc
Location: Nanjing City, Jiangsu Province
Total Area: 52,900 sq.m.
Design Time: October, 2004
Completion : 2007
Co-operation: Architects & Engineers Co., LTD
　　　　　　 of Southeast University

设计构思：

考虑到九龙湖校区总体建筑风格，以及图书馆作为校区中心建筑的需要，在建筑造型上，图书馆对传统文脉、时代精神及文化特征作出了更进一步的探索。首先，建筑采用了传统经典的三段式古典立面形式。其次，建筑采用灰蓝色坡屋顶，以取得与周边院系群及公共教学楼形式上的呼应。同时，建筑又以强烈的虚实对比、体块的穿插产生很强的现代感，形成庄重、严肃的形象特征。在细部设计上，墙体的凹凸产生丰富的阴影，使立面具有强烈的立体感。

东南大学榴园宾馆
LIUYUAN HOTEL OF SOUTHEAST UNIVERSITY

设 计 人：齐康、赵辰、周玉麟	Designer: QI Kang, ZHAO Chen, ZHOU Yu-lin
工程地点：江苏省南京市	Location: Nanjing City, Jiangsu Province
工程规模：11000 平方米	Total Area: 11,000 sq.m.
设计时间：1990–1992	Design Time: 1990-1992
建成时间：1993	Completion: 1993
合作单位：东南大学建筑设计研究院	Co-operation: Architects & Engineers Co., LTD of Southeast University
获奖情况：1994年建设部优秀设计三等奖 1993年江苏省优秀设计一等奖 1995年国家教委优秀建筑设计二等奖	Prize: The Third Prize of Architecture Academic Award, Construction Ministry, 1994 The First Prize of Jiangsu Provincial Architecture Academic Award, 1993 The Second Prize of Architecture Award, National Education Ministry, 1995

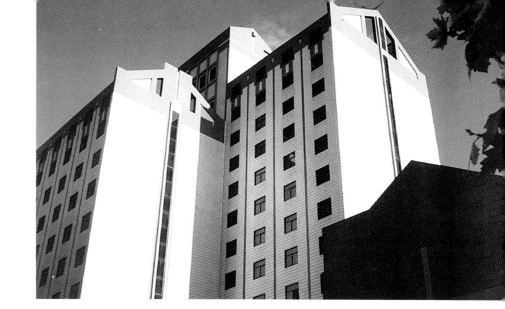

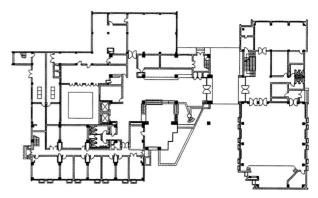

中国科学技术大学生命科学楼
LIFE SCIENCES BUILDING OF USTC

设 计 人：齐康、郑炘、孙曦
工程地点：安徽省合肥市
设计规模：38340 平方米
设计时间：2001.12
建成时间：2003
合作单位：东南大学建筑设计研究院
获奖情况：2005 年教育部优秀建筑设计三等奖

Designer:	QI Kang, ZHENG Xin, SUN Xi
Location:	Hefei City, Anhui Province
Total Area:	38,340 sq.m.
Design Time:	December, 2001
Completion:	2003
Co-operation:	Architects & Engineers Co., LTD of Southeast University
Prize:	The Third Prize of Architecture Academic Award, Education Ministry, 2005

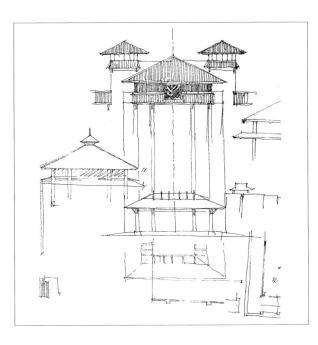

南京农业大学主楼
MAIN BUILDING OF NANJING AGRICULTURAL UNIVERSITY

设 计 人：齐康、张十庆、叶菁、王辉	Designer: QI Kang, ZHANG Shi-qing, YE Jing, WANG Hui
工程地点：江苏省南京市	Location: Nanjing City, Jiangsu Province
工程规模：24190 平方米	Total Area: 24,190 sq.m.
设计时间：1999.7–1999.12	Design Time: July, 1999-December, 1999
建成时间：2000.8	Completion: August, 2000
合作单位：农业部南京设计院	Co-operation: Nanjing Institute of Architectural Design, Agriculture Ministry

南京农业大学逸夫教学楼
YIFU BUILDING OF NANJING AGRICULTURAL UNIVERSITY

设 计 人：齐康、金俊
工程地点：江苏省南京市
工程规模：31850 平方米
设计时间：2002.4
建成时间：2004.
合作单位：农业部南京设计院

Designer: QI Kang, JIN Jun
Location: Nanjing City, Jiangsu Province
Total Area: 31,850 sq.m.
Design Time: April, 2002
Completion: 2004
Co-operation: Nanjing Institute of Agricultural Design, Agriculture Ministry

南京农业大学金陵研究院
JINLING ACADEME OF NANJING AGRICULTURAL UNIVERSITY

设 计 人：齐康、金俊、叶菁	Designer: QI Kang, JIN Jun, YE Jing
工程地点：江苏省南京市	Location: Nanjing City, Jiangsu Province
工程规模：8680 平方米	Total Area: 8,680 sq.m.
设计时间：1997.4–1997.8	Design Time: April, 1997- August, 1997
建成时间：1998.8	Completion : August, 1998
合作单位：东南大学建筑设计研究院	Co-operation: Architects & Engineers Co., LTD of Southeast University

设计构思：

　　金陵研究院建于从北大门开始的校园主轴线终端——田径场的西侧。总体布局采用两个反对称的"L"形平面，并以塔楼为中心，达到均衡。造型设计上，在多处采用了拱廊、坡屋顶等传统语汇的同时，对传统语汇加以变异，进行重组、重构，并采用新技术、新材料，创造出源于传统又有时代气息的新建筑。

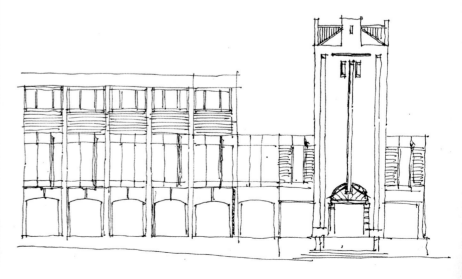

南京农业大学第四教学楼
THE FOURTH ACADEMIC BUILDING OF NANJING AGRICULTURAL UNIVERSITY

设 计 人：齐康、金俊、李烽、齐昉、穆勇
工程地点：江苏省南京市
工程规模：15500 平方米
设计时间：2006.3
建成时间：2008.11
合作单位：东南大学建筑设计研究院

Designer:	QI Kang, JIN Jun, LI Feng, QI Fang, MU Yong
Location:	Nanjing City, Jiangsu Province
Total Area:	15,500 sq.m.
Design Time:	March, 2006
Completion:	November, 2008
Co-operation:	Architects & Engineers Co., LTD of Southeast University

大连理工大学伯川图书馆
BOCHUAN LIBRARY OF DALIAN UNIVERSITY OF TECHNOLOGY

设 计 人：齐康、胡文荟
工程地点：辽宁省大连市
工程规模：20000平方米
设计时间：1996.6–1997.6
建成时间：1998.12
合作单位：大连理工大学建筑设计研究院
获奖情况：2000年教育部优秀工程设计二等奖
　　　　　大连市优秀设计一等奖

Designer:	QI Kang, HU Wen-hui
Location:	Dalian City, Liaoning Province
Total Area:	20,000 sq.m.
Design Time:	June, 1996-June, 1997
Completion:	December, 1998
Co-operation:	Dalian University of Technology Institute of Architectural Design
Prize:	The Second Prize of Architecture Academic Award, Education Ministry, 2000
	The First Prize of Dalian Architecture Academic Award

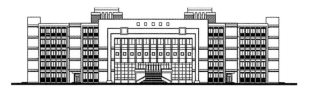

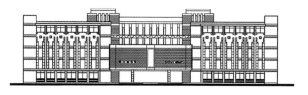

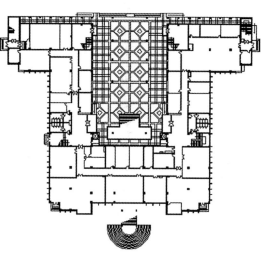

青岛理工大学图书馆
LIBRARY OF QINGDAO TECHNOLOGICAL UNIVERSITY

设 计 人：齐康、马永莉、万正旸
工程地点：山东省青岛市
工程规模：18900 平方米
设计时间：2003
建成时间：2004.10
合作单位：青岛理工大学建筑设计院

Designer: QI Kang, MA Yong-li, WAN Zheng-yang
Location: Qingdao City, Shandong Province
Total Area: 18,900 sq.m.
Design Time: 2003
Completion: October, 2004
Co-operation: Architectural Design Institute of Qingdao Technological University

大连水产学院文夫图书馆
WENFU LIBRARY OF DALIAN FISHERIES COLLEGE

设 计 人：齐康、胡文荟
工程地点：辽宁省大连市
工程规模：13000 平方米
设计时间：2000
建成时间：2002.9
合作单位：大连理工大学建筑设计研究院

Designer: QI Kang, HU Wen-hui
Location: Dalian City, Liaoning Province
Total Area: 13,000 sq.m.
Design Time: 2000
Completion: September, 2002
Co-operation: Dalian University of Technology Institute of Architectural Design

南京航空航天大学
逸夫科技馆

YIFU BUILDING OF SCIENCE & TECHNOLOGY, NANJING UNIVERCITY OF AERONAUTICS

设 计 人：齐康、郑炘
工程地点：江苏省南京市
工程规模：8700平方米
设计时间：1994–1995
建成时间：1996
合作单位：南京航空航天大学建筑设计室
获奖情况：1997年教育部优秀工程设计
　　　　　三等奖
　　　　　1997年教育部邵逸夫先生
　　　　　第七批赠款工程一等奖

Designer: QI Kang, Zheng Xin
Location: Nanjing City, Jiangsu Province
Total Area: 11,000 sq.m.
Design Time: 1994-1995
Completion: 1996
Co-operation: Nanjing University of Aeronautics and Astronautics Architectural Design Studio
Prize: The Third Prize of Architecture Academic Award, Education Ministry, 1997
The First Prize of Architecture Design Award of the Projects of the Seventh Batch Grant Project Invested by Mr. SHAO Yi-fu, 1997

南京航空航天大学综合楼
COMPLEX BUILDING IN NANJING UNIVERSITY OF AERONAUTICS

设 计 人：齐康、郑炘、姜辉、何泰曙
工程地点：江苏省南京市
工程规模：20000 平方米
设计时间：2000.12
建成时间：2004
合作单位：东南大学建筑设计研究院

Designer: QI Kang, ZHENG Xin,
 JIANG Hui, HE Tai-shu
Location: Nanjing City, Jiangsu Province
Total Area: 20,000 sq.m.
Design Time: December, 2000
Completion: 2004
Co-operation: Architects & Engineers Co., LTD
 of Southeast University

中国矿业大学教学楼
MAIN BUILDING OF CHINA UNIVERSITY OF MINING AND TECHNOLOGY

设 计 人：齐康、张宏、邓浩等
工程地点：江苏省徐州市
工程规模：25000 平方米
设计时间：2002
建成时间：2004
合作单位：徐州市第二建筑设计院

Designer: QI Kang, ZHANG Hong, DENG Hao, etc
Location: Xuzhou City, Jiangsu Province
Total Area: 25,000 sq.m.
Design Time: 2002
Completion: 2004
Co-operation: The Second Architectural Design Institute of Xuzhou

徐州师范大学教学主楼群
MAIN BUILDINGS OF XUZHOU NORMAL UNIVERSITY

设 计 人：齐康、齐昉、郑炘、孙磊磊
工程地点：江苏省徐州市
工程规模：46580 平方米
设计时间：2002.5
建成时间：2004
合作单位：东南大学建筑设计研究院

Designer: QI Kang, QI Fang,
             ZHENG Xin, SUN Lei-lei
Location: Xuzhou City, JIangsu Province
Total Area: 46,580 sq.m.
Design Time: May, 2002
Completion: 2004
Co-operation: Architects & Engineers Co., LTD
             of Southeast University

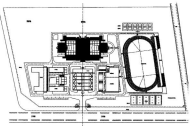
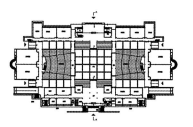

徐州工程学院体育馆
THE GYMNASIUM OF XUZHOU INSTITUTE OF TECHNOLOGY

设 计 人：齐康、张宏等
工程地点：江苏省徐州市
工程规模：8000 平方米
设计时间：1998
建成时间：2000
合作单位：徐州市第二建筑设计院

Designer: QI Kang, ZHANG Hong, etc
Location: Xuzhou City, Jiangsu Province
Total Area: 8,000 sq.m.
Design Time: 1998
Completion: 2000
Co-operation: The Second Institute of Architectural Design of Xuzhou

徐州工程学院图书馆
LIBRARY OF XUZHOU INSTITUTE OF TECHNOLOGY

设 计 人：齐康、张宏等	Designer: QI Kang, ZHANG Hong, etc
工程地点：江苏省徐州市	Location: Xuzhou City, Jiangsu Province
工程规模：18000平方米	Total Area: 18,000 sq.m.
设计时间：1998	Design Time: 1998
建成时间：2000	Completion: 2000
合作单位：徐州市第二建筑设计院	Co-operation: The Second Institute of Architectural Design of Xuzhou

盐城工学院主楼
MAIN BUILDING OF YANCHENG INSTITUE OF TECHNOLOGY

设 计 人：齐康、朱竞翔	Designer: QI Kang, ZHU Jing-xiang
工程地点：江苏省盐城市	Location: Yancheng City, Jiangsu Province
工程规模：9700 平方米	Total Area: 9,700 sq.m.
设计时间：1994–1996	Design Time: 1994-1996
建成时间：1996	Completion: 1996
合作单位：盐城市建筑设计研究院	Co-operation: Yancheng Institute of Architectural Design

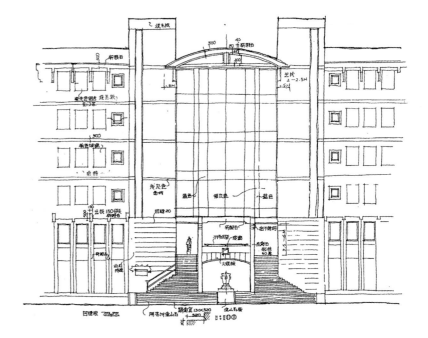

南京鼓楼医院急救中心
EMERGENCY CENTER OF GULOU HOSPITAL, NANJING

设 计 人：齐康、邹式汀、王建国
工程地点：江苏省南京市
工程规模：6000 平方米
设计时间：1988–1991
建成时间：1992
合作单位：南京市建筑设计研究院
获奖情况：1995 年南京市级优秀工程设计三等奖

Designer: QI Kang, ZOU Shi-ting, WANG Jian-guo
Location: Nanjing City, Jiangsu Province
Total Area: 6,000 sq.m.
Design Time: 1988-1991
Completion: 1992
Co-operation: Nanjing Institute of Architectural Design
Prize: The Third Prize of Nanjing Architecture Academic Award, 1995

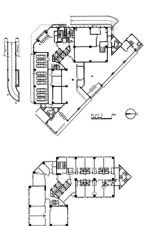

南京鼓楼医院高层病房
HIGH-RISE WARD OF GULOU HOSPITAL, NANJING

设 计 人：齐康、金俊、邹式汀
工程地点：江苏省南京市
工程规模：27680 平方米
设计时间：1997–1999
建成时间：2000
合作单位：南京市建筑设计研究院

Designer:	QI Kang, JIN Jun, ZOU Shi-ting
Location:	Nanjing City, Jiangsu Province
Total Area:	27,680 sq.m.
Design Time:	1997-1999
Completion:	2000
Co-operation:	Nanjing Institute of Architectural Design

河南许昌市政府大楼
CITY HALL OF XUCHANG, HENAN

设 计 人：齐康、潘晓莉
工程地点：河南省许昌市
工程规模：34000 平方米
设计时间：1999
建成时间：2002
合作单位：镇江工业设计院

Designer:	QI Kang, PAN Xiao-li
Location:	Xuchang City, Henan Province
Total Area:	34,000 sq.m.
Design Time:	1999
Completion:	2002
Co-operation:	Zhenjiang Institute of Industrial Design

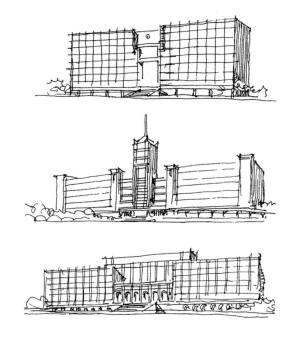

江苏省常熟市政府大楼
CITY HALL OF CHANGSHU, JIANGSU

设 计 人：齐康、王建国、齐昉
工程地点：江苏省常熟市
工程规模：19500 平方米
设计时间：1987
建成时间：1992
合作单位：常熟市建筑设计院
　　　　　东南大学建筑设计研究院

Designer:　　QI Kang, WANG Jian-guo, QI Fang
Location:　　Changshu City, Jiangsu Province
Total Area:　 19,500 sq.m.
Design Time: 1987
Completion:　1992
Co-operation: Changshu Institute of Architectural
　　　　　　　Design
　　　　　　　Architects & Engineers Co., LTD
　　　　　　　of Southeast University

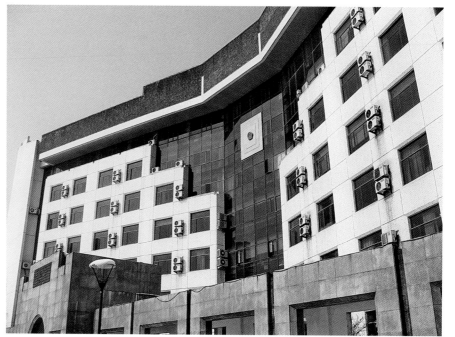

江苏省国家安全教育展览馆
NATIONAL SECURITY EDUCATION EXHIBITION HALL, JIANGSU

设 计 人：齐康、叶菁、张四维、刘媛、朱巍	Designer: QI Kang, YE Jing, ZHANG Si-wei, LIU Yuan, ZHU Wei
工程地点：江苏省南京市	Location: Nanjing City, Jiangsu Province
工程规模：1815 平方米	Total Area: 1,815 sq.m.
设计时间：2004.10	Design Time: October, 2004
建成时间：2008	Completion: 2008
合作单位：东南大学建筑设计研究院	Co-operation: Architects & Engineers Co., LTD of Southeast University

南京市中级人民法院审判庭
NANJING INTERMEDIATE PEOPLE'S COURT ADJUDICATION DIVISION

设 计 人：齐康、倪祥寿
工程地点：江苏省南京市
设计时间：1998
建成时间：2000
合作单位：南京市民用建筑设计院

Designer:　　QI Kang, NI Xiang-shou
Location:　　Nanjing City, Jiangsu Province
Design Time:　1998
Completion:　 2000
Co-operation: Civil Architectural Design Institute of Nanjing

南京市栖霞区法院审判综合楼

COURT DESIGN OF XIXIA DISTRICT, NANJING

设 计 人：齐康、杨志疆、齐昉、
　　　　　沈晓梅、马莉芬
工程地点：江苏省南京市
工程规模：17835 平方米
设计时间：2006.11
建成时间：2009

Designer: QI Kang, YANG Zhi-jiang,
QI Fang, SHEN Xiao-mei,
MA Li-fen
Location: Nanjing City,
Jiangsu Province
Total Area: 17,835 sq.m.
Design Time: November, 2006
Completion: 2009

泉州东湖公园
DONGHU PARK, QUANZHOU

设 计 人：齐康、赖聚奎、张十庆	Designer: QI Kang, LAI Ju-kui, ZHANG Shi-qing
工程地点：福建省泉州市	Location: Quanzhou City, Fujian Province
工程规模：占地 20 公顷	Total Area: 20 hectares
设计时间：1990–1996	Design Time: 1990-1996
建成时间：1997	Completion: 1997
合作单位：泉州市规划设计研究院	Co-operation: Quanzhou City Planning & Design Institute

溧水永寿寺塔塔院
THE COURTYARD OF YONGSHOU PAGODA, LISHUI

设 计 人：齐康、叶菁、孙磊磊
工程地点：江苏省溧水县
工程规模：1140平方米
设计时间：1999.11–2000.6
建成时间：2001.7
合作单位：南京市环都建筑研究院
　　　　　溧水县建筑设计室

Designer: QI Kang, YE Jing, SUN Lei-lei
Location: Lishui County, Jiangsu Province
Total Area: 1,140 sq.m.
Deign Time: November, 1999-June, 2000
Completion: July, 2001
Co-operation: Nanjing Huandu Institute of Architectural Design, Lishui Architectural Studio

镇江碧榆园
BIYU GARDEN OF ZHENJIANG

设 计 人：齐康、陈宗钦、张宁、潘晓莉、李俊霞
工程地点：江苏省镇江市
工程规模：13000 平方米
设计时间：1993-2002
建成时间：2003
合作单位：镇江市工业设计院

Designer: QI Kang, CHEN Zong-qin, ZHANG Ning, PAN Xiao-li, LI Jun-xia
Location: Zhenjiang City, Jiangsu Province
Total Area: 13,000 sq.m.
Design Time: 1993-2002
Completion: 2003
Co-operation: Zhenjiang Institute of Industrial Design

厦门鼓浪屿别墅宾馆
GULANGYU VILLA HOTEL OF XIAMEN

设 计 人：赖聚奎、陈宗钦、齐康
工程地点：福建省厦门市
设计时间：1987
建成时间：1989

Designer: LAI Ju-kui, CHEN Zong-qin, QI Kang
Location: Xiamen City, Fujian Province
Design Time: 1987
Completion: 1989

福建晋江八仙山公园
JINJIANG BAXIANSHAN PARK, FUJIAN

设 计 人：齐康、叶菁、何柯、
　　　　　张成武
工程地点：福建省晋江市
设计时间：2007.7
建成时间：在建
合作单位：东南大学建筑设计研究院

Designer: QI Kang, YE Jing,
　　　　　He Ke, Zhang Cheng-wu
Location: Jinjiang City,
　　　　　Fujian Province
Design Time: July, 2007
Completion: Under Construction
Co-operation: Architects & Engineers
　　　　　Co., LTD of Southeast
　　　　　University

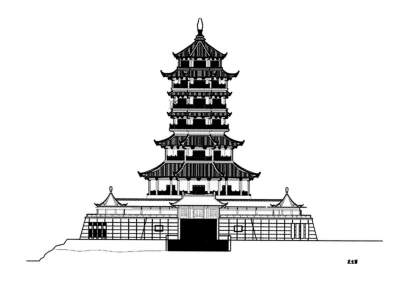

厦门园博园杏林阁
YUANBOYUAN XINGLIN PAVILION IN XIAMEN

设 计 人：齐康、金俊、张成武、朱亚楠	Designer: QI Kang, JIN Jun, ZHANG Cheng-wu, ZHU Ya-nan
工程地点：福建省厦门市	Location: Xiamen City, Fujian Province
工程规模：5291 平方米	Total Area: 5,291 sq.m.
设计时间：2006.7–2007.1	Design Time: July, 2006-January, 2007
建成时间：2007	Completion: September, 2007
合作单位：东南大学建筑设计研究院	Co-operation: Architects & Engineers Co., LTD of Southeast University

张家界国家森林公园入口门票站
ZHANGJIAJIE NATIONAL FOREST PARK ENTRANCE TICKET STATION

设 计 人：齐康、张弦、杨志疆、朱亚楠
工程地点：湖南省张家界市
工程规模：9750 平方米
设计时间：2005.9–2007.9
建成时间：2009.7

Designer:	QI Kang, ZHANG Xian, YANG Zhi-jiang, ZHU Ya-nan
Location:	Zhangjiajie City, Hunan Province
Total Area:	9,750 sq.m.
Design Time:	September, 2005-September, 2007
Completion:	July, 2009

江苏仪化大酒店
YIHUA HOTEL, JIANGSU

设 计 人：齐康、邹式汀、王建国	Designer: QI Kang, ZOU Shi-ting, WANG Jian-guo
工程地点：江苏省扬州市	Location: Yangzhou City, Jiangsu Province
工程规模：17000 平方米	Total Area: 17,000 sq.m.
设计时间：1993-1995	Design Time: 1993-1995
建成时间：1995	Completion: 1995
合作单位：南京市建筑设计研究院	Co-operation: Nanjing Institute of Architectural Design

扬州西园大酒店
XIYUAN HOTEL, YANGZHOU

设 计 人：齐康、段进、许以立	Designer: QI Kang, DUAN Jin, XU Yi-li
工程地点：江苏省扬州市	Location: Yangzhou City, Jiangsu Province
工程规模：38000 平方米	Total Area: 38,000 sq.m.
设计时间：1995–1997	Design Time: 1995-1997
建成时间：1997	Completion: 1997
合作单位：南京市建筑设计研究院	Co-operation: Nanjing Architectural Design Institute
获奖情况：1997 年江苏省优秀工程设计三等奖 1997 年江苏省城乡建设系统优秀工程设计二等奖	Prize: The Third Prize of Jiangsu Provincial Architecture Academic Award, 1997 The Second Prize of Jiangsu Provincial Construction System of Town, 1997

南京雨花台青少年活动中心
NANJING YUHUATAI CENTER FOR YOUTHS

设 计 人：齐康、王彦辉、林艳燕、何柯
工程地点：江苏省南京市
工程规模：1340平方米
设计时间：2008
建成时间：2009.8
合作单位：东南大学建筑设计研究院

Designer: QI Kang, WANG Yan-hui,
              LIN Yan-yan, HE Ke
Location: Nanjing City, Jiangsu Province
Total Area: 1,340 sq.m.
Design Time: August, 2008
Completion: Under Construction
Co-operation: Architects & Engineers Co., LTD
              of Southeast University

江苏省钟山干部疗养院
ZHONGSHAN SANATORIUM OF JIANGSU PROVINCE

设 计 人：齐康、张宏、马晓东	Designer: QI Kang, ZHANG Hong, MA Xiao-dong
工程地点：江苏省南京市	Location: Nanjing City, Jiangsu Province
工程规模：4000 平方米	Total Area: 4,000 sq.m.
设计时间：1997–1998	Design Time: 1997-1998
建成时间：1999	Completion: 1999
合作单位：东南大学建筑设计研究院	Co-operation: Architects & Engineers Co., LTD of Southeast University

镇江市福利院
ZHENJIANG WELFARE HOSPITAL

设 计 人：齐康、王彦辉、潘晓莉
工程地点：江苏省镇江市
工程规模：26000平方米
设计时间：1995-1997
建成时间：1998
合作单位：镇江市工业设计院

Designer: QI Kang, WANG Yan-hui, PAN Xiao-li
Location: Zhenjiang City, Jiangsu Province
Total Area: 26,000 sq.m.
Design Time: 1995-1997
Completion: 1998
Co-operation: Zhenjiang Institute of Industrial Design

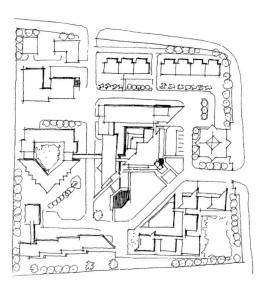

中国科学院土壤研究所办公楼
BUILDING OF SOIL RESEARCH INSTITUTE, CHINESE ACADEMY OF SCIENCES

设 计 人：齐康、齐昉、王印武
工程地点：江苏省南京市
工程规模：9700 平方米
设计时间：1994–1996
建成时间：1996
合作单位：东南大学建筑设计研究院

Designer:	QI Kang, QI Fang, WANG Yin-wu
Location:	Nanjing City, Jiangsu Province
Total Area:	9,700 sq.m.
Design Time:	1994-1996
Completion:	1996
Co-operation:	Architects & Engineers Co., LTD of Southeast University

中国科学院澄江古生物研究站
CHENGJIANG PALEOBIOLOGY ACADEME, YUNNAN

设 计 人：齐康、华峰、王莉
工程地点：云南省澄江县
工程规模：32000 平方米
设计时间：1996–1998
建成时间：1999
合作单位：云南工业大学建筑设计研究院

Designer:	QI Kang, HUA Feng,
		WANG Li
Location:	Chengjiang City, Yunnan Province
Total Area:	32,000 sq.m.
Design Time:	1996-1998
Completion:	1999
Co-operation:	Institute of Architectural Design,
		Yunnan Industrial University

黄山市电业局调度楼
ELECTRIC POWER SCHEDULIND OF HUANGSHAN CITY

设 计 人：陈宗钦、郑炘、齐康
工程地点：安徽省黄山市
工程规模：13000 平方米
设计时间：1998
建成时间：2000
合作单位：镇江工业设计院

Designer:　　CHEN Zong-qin, ZHENG Xin, QI Kang
Location:　　 Huangshan City, Anhui Province
Total Area:　 13,000 sq.m.
Design Time: 1998
Completion:　2000
Co-operation: Zhenjiang Institute of Industrial Design

深圳贝岭居
BEILING BUILDING, SHENZHEN

设 计 人：齐康、赖聚奎
工程地点：广东省深圳市
工程规模：11515 平方米
设计时间：1990
建成时间：1992

Designer: QI Kang, LAI Ju-kui
Location: Shenzhen City, Guangdong Province
Total Area: 11,515 sq.m.
Design Time: 1990
Completion: 1992

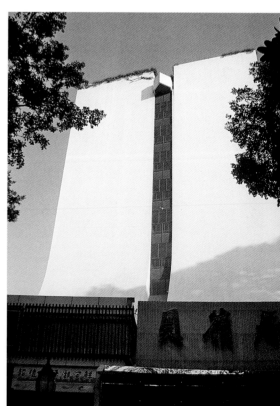

厦门市文联大楼
CULTURE BUILDING OF XIAMEN

设 计 人：齐康、赵晓波、黄印武
工程地点：福建省厦门市
工程规模：7500 平方米
设计时间：1997–1998
建成时间：2000
合作单位：厦门市建筑设计研究院

Designer:	QI Kang, ZHAO Xiao-bo, HUANG Yin-wu
Location:	Xiamen City, Fujian Province
Total Area:	7,500 sq.m.
Design Time:	1997-1998
Completion:	2000
Co-operation:	Architectural Design & Research Institute of Xiamen

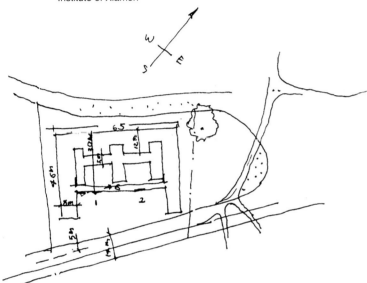
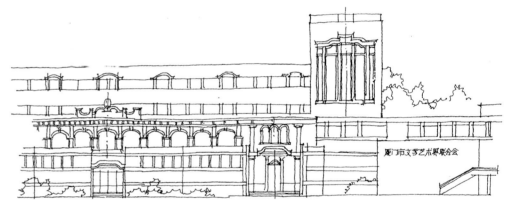

镇江市人防办防空指挥中心
COMMANDING CENTER OF CIVIL AERIAL DEFENCE IN ZHENJIANG

设 计 人：张宏、张振辉
工程地点：江苏省镇江市
工程规模：5600平方米
设计时间：2003
建成时间：2005
合作单位：东南大学建筑设计研究院
获奖情况：2007年南京市优秀建筑设计二等奖

Designer: ZHANG Hong, ZHANG Zhen-hui
Location: Zhenjiang City, Jiangsu Province
Total Area: 5,600 sq.m.
Design Time: 2003
Completion: 2005
Co-operation: Architects & Engineers Co., LTD of Southeast University
Prize: The Second Prize of Nanjing Architecture Academic Award, 2007

中山大学珠海分校伍舜德学术交流中心
WUSHUNDE SCIENTIFIC COMMUNICATION CENTER IN ZHONGSHAN UNIVERSITY

设 计 人：齐康、尹兆俊、季宾
工程地点：广东省珠海市
工程规模：15000平方米
设计时间：2002.10–2003.4
建成时间：2004.10
合作单位：南京金海设计工程有限公司珠海分公司

Designer: QI Kang, YIN Zhao-jun, JI Bin
Location: Zhuhai City, Guangdong Province
Total Area: 15,000 sq.m.
Design Time: October, 2002-April, 2003
Completion: October, 2004
Co-operation: Nanjing Jinhai Design Engineering Co., Ltd., Zhuhai Branch

南京大厂中学
DACHANG MIDDLE SCHOOL, NANJING

设 计 人：齐康、应兆金
工程地点：江苏省南京市
工程规模：5080平方米
设计时间：1989–1990
建成时间：1992
合作单位：东南大学建筑设计研究院

Designer: QI Kang, YING Zhao-jin
Location: Nanjing City, Jiangsu Province
Total Area: 5,080 sq.m.
Design Time: 1989-1990
Completion: 1992
Co-operation: Architects & Engineers Co., LTD of Southeast University

扬州历史纪念碑
YANGZHOU HISTORIC MONUMENTS

设 计 人：齐康、温洋	Designer: QI Kang, WEN Yang
工程地点：江苏省扬州市	Location: Yangzhou City, Jiangsu Province
设计时间：2005	Design Time: 2005
建成时间：2007	Completion: 2007
合作单位：东南大学建筑设计研究院	Co-operation: Architects & Engineers Co., LTD of Southeast University

邳州公园大门
GATE OF PEIZHOU GARDEN

设 计 人：张宏
工程地点：江苏省邳州市
设计时间：1997
建成时间：1997
合作单位：徐州市第二建筑设计院

Designer: ZHANG Hong
Location: Peizhou City, Jiangsu Province
Design Time: 1997
Completion: 1997
Co-operation: The Second Institute of Architectural Design, Xuzhou

南通烈士陵园纪念碑
NANTONG MARTYRS CEMETERY MONUMENTS

设 计 人：齐康
工程地点：江苏省南通市
设计时间：1984
建成时间：1985

Designer: QI Kang
Location: Nantong City, Jiangsu Province
Design Time: 1984
Completion: 1985

南京金陵中学一百周年纪念碑
THE MONUMENT FOR 100TH ANNIVERSARY JINLING SENIOR SCHOOL, NANJING

设 计 人：齐康	Designer: QI Kang
工程地点：江苏省南京市	Location: Nanjing City, Jiangsu Province
设计时间：1988	Design Time: 1988
建成时间：1988	Completion: 1988

缙云县洋潭头后陈山公园景观塔
JINYUN YANGTANTOU LANDSCAPE TOWER IN HOUCHENSHAN PARK

设 计 人：杨志疆、齐康、杨程、李峰、朱巍、钱玉斋	Designer: YANG Zhi-jiang, QI Kang, YANG Cheng, LI Feng, ZHU Wei, QIAN Yu-zhai
工程地点：浙江省缙云县	Location: Jinyun City, Zhejiang Province
工程规模：3234 平方米	Total Area: 3,234 sq.m.
设计时间：2007–2008	Design Time: 2007-2008
建成时间：2012	Completion: 2012

江苏如皋红十四军纪念馆
RUGAO FOURTEEN RED ARMY MEMORIAL, JIANGSU

设 计 人：齐康、朱亚楠
工程地点：江苏省如皋市
工程规模：7715 平方米
设计时间：2008.8
建成时间：2010

Designer: QI Kang, ZHU Ya-nan
Location: Rugao City, Jiangsu Province
Total Area: 7,715 sq.m.
Design Time: August, 2008
Completion: 2010

南昌八大山人纪念馆改扩建
BADASHANREN MEMORIAL HALL RENOVATION AND EXPANSION

设 计 人：齐康、金俊、徐旺
工程地点：江西省南昌市
工程规模：4212 平方米
设计时间：2009.7
建成时间：2012
合作单位：东南大学建筑设计研究院

Designer: QI Kang, JIN Jun, XU Wang
Location: Nanchang City, Jiangxi Province
Total Area: 4,212 sq.m.
Design Time: July, 2009
Completion: 2012
Co-operation: Architects & Engineers Co., LTD of Southeast University

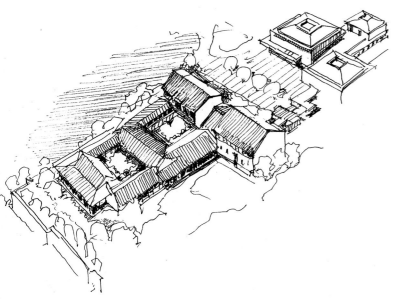

淮北革命根据地纪念馆、纪念碑
HUAIBEI REVOLUTIONARY BASE MEMORIAL, MONUMENT

设 计 人：齐康、金俊、周妍琳、朱亚楠	Designer: QI Kang, JIN Jun, ZHOU Yan-lin, ZHU Ya-nan
工程地点：江苏省泗洪县	Location: Sihong County, Jiangsu Province
工程规模：2450平方米	Total Area: 2,450 sq.m.
设计时间：2009.3	Design Time: March, 2009
建成时间：2012	Completion: 2012

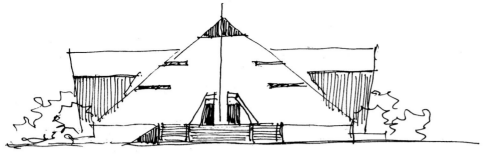

淮安八十二烈士纪念馆
EIGHTY-TWO MARTYRS MEMORIAL, HUAI'AN

设 计 人：	齐康、齐昉、蒋莉等	Designer:	QI Kang, QI Fang, JIANG Li, etc
工程地点：	江苏省淮安市	Location:	Huan'an City, Jiangsu Province
工程规模：	3840 平方米	Total Area:	3840 sq.m.
设计时间：	2009	Design Time:	2009
建成时间：	2010	Completion:	2010
合作单位：	东南大学建筑设计研究院有限公司	Co-operate:	Architects & Engineers Co., LTD of Southeast University

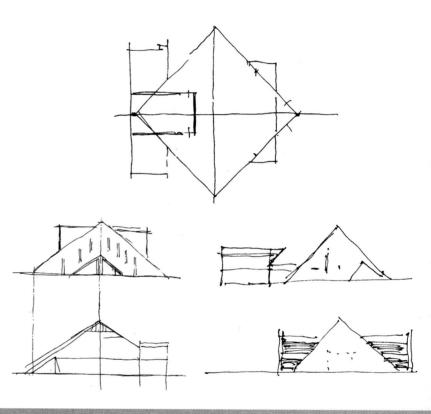

淮安母爱公园爱心塔
LOVE TOWER DESIGN, HUAI'AN

设　计　人：齐康、叶菁、夏明明等
工程地点：江苏省淮安市
工程规模：4880平方米
设计时间：2010–2011
建成时间：2012
合作单位：东南大学建筑设计研究院有限公司

Designer: QI Kang, YE Jing, XIA Ming-ming
Location: Huan'an City, Jiangsu Province
Total Area: 4880 sq.m.
Design Time: 2010-2011
Completion: 2012
Co-operate: Architects & Engineers Co., LTD of Southeast University

淮安感恩广场
ARCHITECTURAL DESIGN OF GAN'EN HALL, HUAI'AN

设 计 人：齐康、叶菁、白鹭飞等
工程地点：江苏省淮安市
工程规模：2916平方米
设计时间：2012–2013
建成时间：2014
合作单位：东南大学建筑设计研究院有限公司

Designer: QI Kang, YE Jing, BAI Lu-fei
Location: Huan'an City, Jiangsu Province
Total Area: 2916 sq.m.
Design Time: 2012-2013
Completion: 2014
Co-operate: Architects & Engineers Co., LTD of Southeast University

重庆杨闇公烈士陵园规划及建筑单体
YANG AN GONG MARTYRS CEMETERY, CHONGQING

设 计 人：齐康、许彬彬等
工程地点：重庆市
工程规模：18720 平方米（用地）
设计时间：2009
建成时间：2011
合作单位：东南大学建筑设计研究院有限公司

Designer:	QI Kang, XU Bin-bin
Location:	Chongqing City
Total Area:	18720 sq.m. (Land Area)
Design Time:	2009
Completion:	2011
Co-operate:	Architects & Engineers of Southeast University

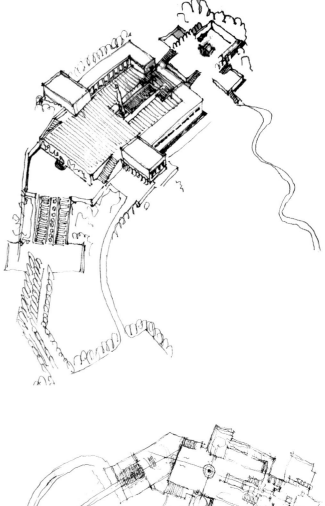

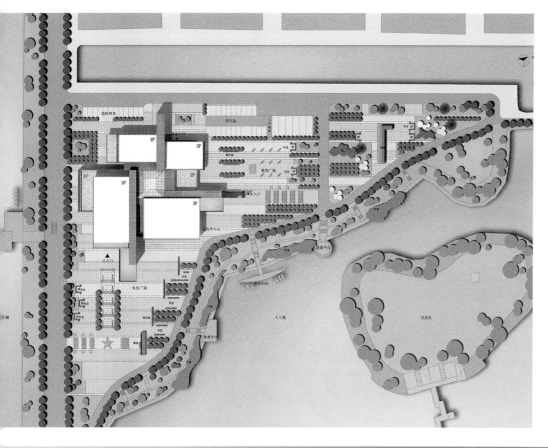

南通市中小学生素质教育基地展览馆
（南通苏中四分区抗日反清乡斗争纪念馆）
PRIMARY AND MIDDLE SCHOOL-STUDENTS QUALITY EDUCATION EXHIBITION HALL IN NANTONG

设 计 人：	齐康、张四维、宫聪、刘九三、张弦	Designer:	Qi Kang, Zhang Si-wei, Gong Cong, Liu Jiu-san, Zhang Xian
工程地点：	江苏省南通市如东县	Location:	Rudong County, Nantong, Jiangsu Province
工程规模：	20915平方米	Total Area:	20915sq.m.
设计时间：	2013–2014	Design Time:	2013-2014
建成时间：	2015	Completion:	2015
合作单位：	东南大学建筑设计研究院有限公司	Co-operate:	Architects & Engineers Co., LTD of Southeast University

设计构思：

本方案通过建筑语言表达了"党的凝聚力"这一主题。体块的叠加正好反映出共产党人当时团结互助，英勇不屈的精神力量。方形的体块在呼应纪念馆建筑肃穆的性质同时，与主题相符。三个向外"冲出"的长方形体块可以象征当时共产党人和人民群众在反清乡运动中冲出敌人用竹篱笆围起来的包围圈，体现了其不屈不挠，英勇抗战的精神。材质用石材干挂，更符合纪念馆凝重沉稳的特质。

考虑到基地周围优美的环境，在景观面较好的方向设计三条片墙，继而形成三块交通盒。依附三片"墙"布置展览空间，二层与一层交错布置，留出来的多余空间底层可以插入水面，顶层形成屋顶平台。屋顶平台可以让浏览的人更好的欣赏周边的美景。在围合的展厅中间设计一块两层通高的玻璃中庭，中庭一层和二层均有围廊，部分围廊用片墙遮挡在其后面，可以形成通透的趣味空间。再现当时反清乡斗争的凶险。中庭顶端是一个攒尖玻璃顶，仿佛将四个体块拉起来，呼应了建筑概念主题。

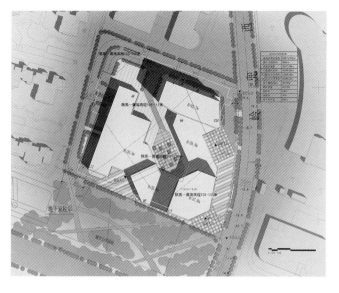

一层平面图

厦门思明研发产业园
SINGMING R & D INDUSTRIAL PARK IN XIAMEN

设 计 人：齐康、张四维、宫聪、王洲、张胜亚
工程地点：福建省厦门市
工程规模：76477 平方米
设计时间：2014
建成时间：2017
合作单位：东南大学建筑设计研究院有限公司

Designer: Qi Kang, Zhang Si-wei, Gong Cong, Wang Zhou, Zhang Sheng-ya
Location: Xiamen, Fujian Province
Total Area: 20915sq.m.
Design Time: 2013-2014
Completion: 2017
Co-operate: Architects & Engineers Co., LTD of Southeast University

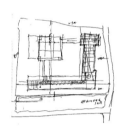
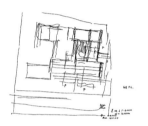

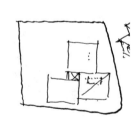

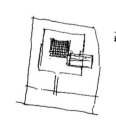
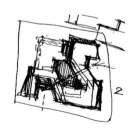

南通汽车客运东站
EASTERN AUTOMOBILE PASSENGER STSTION, NANTONG

设 计 人：齐康、王彦辉、赵倩、高晓明
工程地点：江苏省南通市
工程规模：71000平方米
设计时间：2010–2011
建成时间：2013
合作单位：东南大学建筑设计研究院有限公司
获奖情况：2015年度江苏省城乡建设系统
　　　　　优秀勘察设计建筑设计三等奖

Designer: Qi Kang, Wang Yan-hui, Zhao Qian, Gao Xiao-ming
Location: Nantong City, Jiangsu Province
Total Area: 71000 sq.m.
Design Time: 2010-2011
Completion: 2013
Co-operate: Architects & Engineers Co., LTD of Southeast University
Prize: The Third Prize of Architecture Academic Award 2015, Construotion Department of Jiang Su Province

姜堰博物馆
MUSEUM, JIANGYAN

设 计 人：齐康、王彦辉、张弦、张芳
工程地点：江苏省扬州市姜堰区
工程规模：5338 平方米
设计时间：2011-2012
建成时间：2014
合作单位：东南大学建筑设计研究院
　　　　　有限公司
获奖情况：2015年度教育部优秀勘察设计
　　　　　二等奖
　　　　　2015年度全国优秀工程勘察设
　　　　　计行业奖建筑工程设计项目三
　　　　　等奖

Designer:	Qi Kang, Wang Yan-hui, Zhang Xian, Zhang Fang
Location:	JiangYan District, Yangzhou City, Jiangsu Province
Total Area:	5338 sq.m.
Design Time:	2011-2012
Completion:	2014
Co-operate:	Architects & Engineers Co., LTD of Southeast University
Prize:	The Second Prize of Architecture Academic Award 2015, Education Ministry
The Third Prize of Architecture Academic Award 2015, Construotion Ministry

南京丁山宾馆
DINGSHAN HOTEL, NANJING

设 计 人：齐康、赵国权等
工程地点：南京市
设计时间：1975
建成时间：1976
合作单位：南京市建筑设计研究院

Designer: QI Kang, Zhao Guo-quan
Location: Nanjing City, Jiangsu Province
Design Time: 1975
Completion: 1976
Co-operate: Nanjing Institute of Architecture Design

福州市马尾区图书馆
THE LIBRARY IN MAWEI OF FUZHOU

设 计 人：齐康、金俊、林卫宁、朱晶秋、
　　　　　汪亮、冯然、葛晓峰
工程地点：福建省福州市
工程规模：15000 平方米
建成时间：2016.10

Designer: QI Kang, Jin Jing, Lin Weining,
Zhu Jingqiu, Wang Liang,
Feng Ran, Ge Xiaofeng
Location: Fuzhou City, Fujian Province
Total Area: 15000 sq.m.
Completion: 2016.10

设计构思：

　　图书馆设计方案以历史文化为依托，通过现代科技建造手段及当代前沿的建筑语言，力图展示历史积淀中升华的现代科技魅力。本馆建设摒弃传统图书馆固有的布局模式，为读者创造出全新的家居式阅读体验，在细节设计上着力营造出亲民、温馨、舒适、开放的阅读环境。新馆的服务理念与国际一流图书馆相接轨，由传统的"书本位"提升为"人本位"，并将此理念融入空间布局设计中。

江苏省金坛茅山气象综合实验基地
METEOROLOGICAL COMPREHENSIVE EXPERIMENTAL BASE

设 计 人：齐康、张青萍、金俊、林冀闽、马妍、朱晶秋、许彬彬
工程地点：江苏省常州市
工程规模：6328.6 平方米
建成时间：2014.5

Designer: QI Kang, Zhang Qing-ping, Jin Jing, Lin Ji-min, Ma Yan, Zhu Jing-qiu, Xu Bing-bing
Location: Changzhou City, Jiangsu Province
Total Area: 6328.6 sq.m.
Completion: 2014.5

设计构思：

总体形态上根据茅山自然环境特色，以保持现有原生地貌为宗旨，充分利用地形，将建筑物设计成依山就势的自然摆布形态，最大程度地保护自然环境，同时也使建筑成为环境中的点睛之笔，从而提升整体环境的美感。

建筑设计上充分考虑了茅山自然地貌和文化背景，建筑物单体均采用坡屋顶，既保证了良好的排水和保暖效果，同时错落有致的屋顶有与高山祈福的山势融为一体，同时大飘窗、观景台和观景廊等建筑元素将户外美景更好地引入到建筑的内部。

摄影：贾方

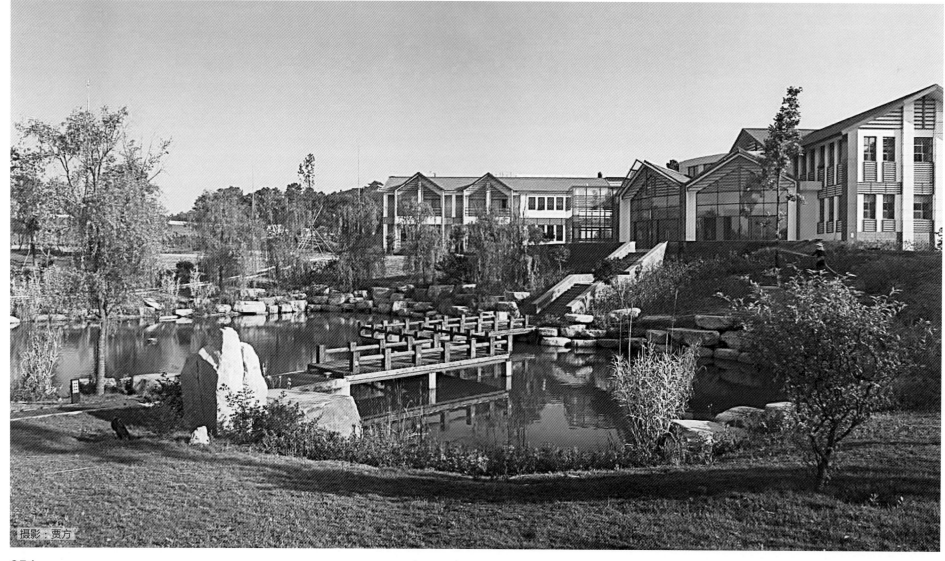
摄影：贾方

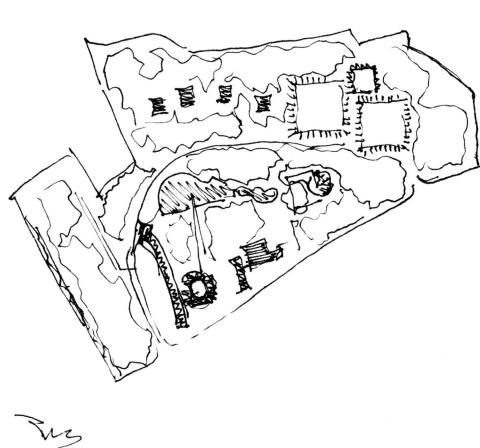

宣城市宛陵湖配套商业服务建筑
COMMERCIAL SERVICES ARCHITECTURE IN XUANCHENG

设 计 人：齐康、金俊、张四维、寿刚、 　　　　　王雪寅、崔毅雄	Designer: QI Kang, Jin Jing, Zhang Siwei, Shou Gang, Wang Xueyin, Cui Yixiong
工程地点：安徽省宣城市	Location: Xuancheng City, Anhui Province
工程规模：33207平方米	Total Area: 33207 sq.m.
建成时间：2015.12	Completion: 2015.12

设计构思：

项目周边现有的自然景观和已有的总规规划都是本方案设计着重考虑的因素，尊重周围人文与自然环境，设计时应充分注意与周边已有的展览馆与大剧院和谐呼应，成为融入宛陵湖景区的建筑群。

本次设计摒弃了低廉糙杂的日用百货批发市场模式，旨在创造一个生态丰富、有主题景观、充满假日气氛的现代商业建筑群，创造一组空间丰富、尺度宜人、环境精致的中高档休闲娱乐场所，使其成为宣城的名片。

南京大学东南楼
SOUTHEAST BUILDING OF NANJING UNIVERSITY

设 计 人：杨廷宝、齐康等
工程地点：江苏省南京市
设计时间：1953
建成时间：1954 年 7 月

Designer: YANG Ting-bao, QI Kang
Location: Nanjing City,
 Jiangsu Province
Design Time: 1953
Completion: July, 1954

泰州中学老校区图书馆
LIBRARY OF TAIZHOU HIGH SHOOL

设 计 人：齐康、王彦辉、叶菁等
工程地点：江苏省泰州市
工程规模：3290平方米
设计时间：2012–2013
建成时间：2014
合作单位：东南大学建筑设计研究院有限公司

Designer:	QI Kang, WANG Yan-hui, YE Jing
Location:	Taizhou City, Jiangsu Province
Total Area:	3290 sq.m.
Design Time:	2012-2013
Completion:	2014
Co-operate:	Architects & Engineers Co., LTD of Southeast University

南京五台山体育馆
WU TAI SHAN STADIUM, NANJING

设 计 人：杨廷宝、齐康、钟训正等	Designer: YANG Tingbao, QI Kang, Zhong Xunzheng
工程地点：江苏省南京市	Location: Nanjing City, Jiangsu Province
工程规模：17930平方米	Total Area: 17930 sq.m.
设计时间：1973	Design Time: 1973
建成时间：1975	Completion: 1975
合作单位：江苏省建筑设计研究院	Co-operate: Jingsu Institute of Architecture Design

设计构思：

南京五台山体育馆位于南京市中心的五台山体育中心西侧，交通便利、闹中取静、环境幽雅。平面为长八角形，南北长88.7米，东西宽76.8米，檐高25.2米。立面造型与结构紧密结合，四周由64根桩柱植根于基岩之中，屋架采用平板型双层三向球结点钢管网架结构，当年就地焊结，用48部卷扬机整体吊装一次性成功，看台采用钢筋混凝土框架结构，柱、梁、板预制与整浇相结合。可举行篮球、排球、乒乓球、羽毛球等球类和体操、技巧、武术、柔道、摔跤、举重、击剑等项目比赛，举办歌舞、杂技、魔术等文艺表演，组织大型庆典、集会、洽谈会、展销等活动。

锦溪人民医院（老年护理院）一期
JINXI PEOPLE'S HOSPITAL, KUNSHAN

设 计 人：齐康、周颖等	Designer: QI Kang, Zhou Ying
工程地点：江苏省昆山市	Location: Kunshan City, Jiangsu Province
工程规模：25473平方米	Total Area: 25473 sq.m.
设计时间：2010.5–2011.8	Design Time: May, 2010-August, 2011
建成时间：2014.5	Completion: May, 2014
合作单位：东南大学建筑设计研究院有限公司	Co-operate: Architects & Engineers Co., LTD of Southeast University
获奖情况：2016年江苏省城乡建设系统优秀勘察设计三等奖	Prize: The Third Prize of Jiangsu Provincial Architecture Academic Award, 2016

设计构思：

　　锦溪人民医院采用半集中式半分散式布局的形式，在获得较高运营效率的同时保证较高的舒适度、自然采光和自然通风。分别设置8个出入口，有效地分解了每天门诊、发热和肠道门诊、急诊、住院及探视、儿童保健、工作人员的人流、货运物流和污物出入口。

　　建筑造型设计借助于墙体的凹凸及坡屋面的设置，将该大体量的建筑物分解为若干个小体量的组合，从而接近了江南民居的宜人尺度。建筑以白色和灰色为主色调，以获得"粉墙黛瓦"的艺术效果。同时，借用江南园林中常用的对景的设计手法，在门诊入口正对面设置中心水景庭院，种植了当地特产并蒂莲。并通过圆洞门的设置，使患者有置身于江南园林之感。该中心庭院有助于患者能快速正确地辨认当前位置并明确方向，不仅能增强患者的安心感，还会大大提高医院的交通效率。

蚌埠规划勘测研究中心设计方案
ARCHITECTURE DESIGN FOR THE PLAN & SURVEY CENTER OF BENGBU

设 计 人：齐康、李烽、马莉芬	Designer: QI Kang, LI Feng, MA Li-fen
工程地点：安徽省蚌埠市	Location: Bengbu City, Anhui Province
工程规模：12505平方米	Total Area: 12,505 sq.m.
设计时间：2007.8	Design Time: August, 2007
建成时间：在建	Completion: Under Construction
合作单位：蚌埠市规划设计研究院	Co-operate: Bengbu Academy of Urban Planning & Design

大连甘井子区图书档案馆设计方案
GANJINGZI LIBRARY & ARCHIVES CENTER, DALIAN

设 计 人：齐康、范双丹、李烽	Designer: QI Kang, FAN Shuang-dan, LI Feng
工程地点：辽宁省大连市	Location: Dalian City, Liaoning Province
工程规模：22457平方米	Total Area: 22,457 sq.m.
设计时间：2007.8	Design Time: August, 2007
建成时间：在建	Completion: Under Construction

禹州市钧窑遗址博物馆设计方案
YUZHOU JUN PORCELAIN MUSEUM DESIGN

设 计 人：齐康、张弦、马莉芬、潘晓莉
工程地点：河南省禹州市
工程规模：9553 平方米
设计时间：2006.3
建成时间：在建

Designer: QI Kang, ZHANG Xian, MA Li-fen, PAN Xiao-li
Location: Yuzhou City, Henan Province
Total Area: 9,553 sq.m.
Design Time: March, 2006
Completion: Under Construction

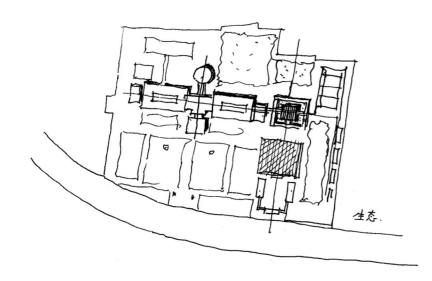

青海湖旅游接待中心设计方案
QINGHAI LAKE TOURISM RECEPTION CENTER DESIGN

设 计 人：齐康、王林、王小嘉	Designer: QI Kang, WANG Lin, WANG Xiao-jia
工程地点：青海省	Location: Qinghai Province
工程规模：2170平方米	Total Area: 2,170 sq.m.
设计时间：2003.12	Design Time: December, 2003
建成时间：在建	Completion: Under Construction

微山县文化艺术中心设计方案
CULTURAL ARTS CENTER DESIGN OF WEISHAN COUNTY

设 计 人：齐康、李晓雪、高晓明、林艳燕	Designer: QI Kang, LI Xiao-xue, GAO Xiao-ming, LIN Yan-yan
工程地点：山东省微山县	Location: Weishan County, Shandong Province
工程规模：637 亩	Total Area: 637 mu
设计时间：2008.6	Design Time: June, 2008
建成时间：在建	Completion: Under Construction

大连普兰店市青少年活动中心设计方案
DESIGN OF PULANDIAN CENTER FOR YOUNG PEOPLE, DALIAN

设 计 人：齐康、何妍亭	Designer: QI Kang, HE Yan-ting
工程地点：辽宁省普兰店市	Location: Pulandian City, Liaoning Province
工程规模：24300 平方米	Total Area: 24,300 sq.m.
设计时间：2008.8	Design Time: August, 2008
建成时间：在建	Co-operation: Under Construction

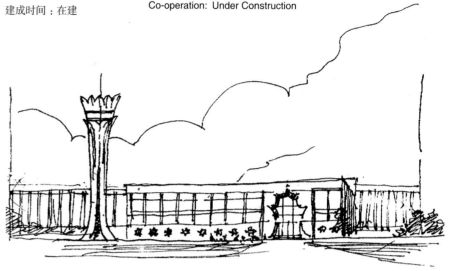

青州博物馆设计方案
QINGZHOU MUSEUM DESIGN

设 计 人：齐康、叶菁、张芳、宋晶秋
工程地点：山东省青州市
工程规模：1340 平方米
设计时间：2007.10
建成时间：在建
合作单位：东南大学建筑设计研究院

Designer: QI Kang, YE Jing,
             ZHANG Fang, SONG Jin-qiu
Location: Qingzhou City,
             Shandong Province
Total Area: 1,340 sq.m.
Design Time: October, 2007
Completion: Under Construction
Co-operation: Architects & Engineers Co.,
              LTD of Southeast University

267

北京某区城市设计方案
URBAN PLANING OF BEIJING LOCAL DISTRICT

设 计 人：齐康、杨志疆、王宇、李芝也、高晓明
工程地点：北京市
工程规模：19.58公顷
设计时间：2009.5
建成时间：在建

Designer: QI Kang, Yang Zhi-jiang, WANG Yu, LI Zhi-ye, GAO Xiao-ming
Location: Beijing City
Total Area: 19.58 hectares
Design Time: May, 2009
Completion: Under Construction

南京农业大学理科实验楼设计方案
SCIENCE LABORATORY BUILDING OF NANJING AGRICULTURAL UNIVERSITY

设 计 人：齐康、齐昉、金俊、李烽、
　　　　　马莉芬、徐旺、庞博
工程地点：江苏省南京市
工程规模：42464 平方米
设计时间：2008.10
建成时间：在建
合作单位：东南大学建筑设计研究院

Designer: QI Kang, QI Fang, JIN Jun, LI Feng,
　　　　　MA Li-fen, XU Wang, PANG Bo
Location: Nanjing City, Jiangsu Province
Total Area: 42,464 sq.m.
Design Time: October, 2008
Completion: Under Construction
Co-operation: Architects & Engineers Co., LTD
　　　　　　　of Southeast University

北京市通州某地段规划设计方案
PLANNING AND DESIGN OF LOCAL TOWN IN TONGZHOU, BEIJING

设 计 人：齐康、王宇
工程地点：北京市通州区
工程规模：8.69公顷
设计时间：2009.6
建成时间：在建

Designer: QI Kang, Wang Yu
Location: Tongzhou District, Beijing City
Total Area: 8.69 hectares
Design Time: June, 2009
Completion: Under Construction

镇江碧榆园国际会议中心设计方案
ZHENJIANG BIYUYUAN INTERNATIONAL CONFERENCE CENTRE DESIGN

设 计 人：齐康、王彦辉、汪亮、冯然
工程地点：江苏省镇江市
工程规模：2700平方米
设计时间：2009
建成时间：在建

Designer: QI Kang, WANG Yan-hui,
              WANG Liang, FENG Ran
Location: Zhenjiang City, Jiangsu Province
Total Area: 2,700 sq.m.
Design Time: 2009
Completion: Under Construction

南京浦镇车辆有限公司 1915 公馆设计方案
PUZHEN VEHICLES CO., LTD. 1915 RESIDENCE, NANJING

设　计　人：齐康、韦庚男、何柯、许彬彬
工程地点：江苏省南京市
工程规模：4150 平方米
设计时间：2009.5
建成时间：在建

Designer: QI Kang, WEI Geng-nan, HE Ke, XU Bin-bin
Location: Nanjing City, Jiangsu Province
Total Area: 4,150 sq.m.
Design Time: May, 2009
Completion: Under Construction

厦门市同安区佛心寺设计方案
FOXIN TEMPLE IN TONGAN DISTRICT, XIAMEN

设 计 人：齐康、金俊、朱晶秋、雍玉洁
工程地点：福建省厦门市同安区
工程规模：8500 平方米
设计时间：2010.5–2012.6

Designer: QI Kang, JIN Jun, ZHU Jin-qiu, YONG Yu-jie
Location: Tongan District, Xiamen City, Fujian Province
Total Area: 8500 sq.m.
Design Time: 2010.5-2012.6

设计构思：

佛心寺位于厦门市同安区。佛心寺整体按照轴线布局。从南往北依次为放生池、天王殿、大雄宝殿、藏经阁。藏经阁呈U型布局，层层退台，环抱大雄宝殿。

新一代天气雷达建设项目（天语舟）设计方案

A PROJECTION OF NEW GENERATION WEATHER RADAR CONSTRCTION (TIANYU SHIP)

设 计 人：齐康、金俊、寿刚、仲早立、胡长涓、陈小坚
工程地点：福建省厦门市
工程规模：5805 平方米
设计时间：2014.6–2015.10
建成时间：2017
合作单位：东南大学建筑设计研究院有限公司

Designer: QI Kang, JIN Jun, SHOU Gang, ZHONG Zao-li, HU Chang-juan, CHEN Xiao-jian
Location: Xiamen City, Fujian Province
Total Area: 5805 sq.m.
Design Time: June, 2014-October, 2015
Completion: 2017
Co-operation: Architects & Engineers Co., LTD of Southeast University

设计构思：

"天语舟"位于福建省厦门市海沧区蔡尖尾山南麓。从空中俯瞰"天语舟"会觉得它像中国古代的司南，又好似一艘自天外驾云渡水而来的船。它的设计表达了天人融合的理念，强调人与自然和谐对话。

义乌细菌战受难同胞纪念馆设计方案
MEMORIAL OF THE VICTIMS IN YIWU GERM WARFARE

设 计 人：齐康、金俊、寿刚、张胜亚、沈骁茜、展泽励
工程地点：浙江省义乌市
工程规模：9000 平方米
设计时间：2014.9–2015.12
合作单位：东南大学建筑设计研究院有限公司

Designer: QI Kang, JIN Jun, SHOU Gang, ZHANG Sheng-ya, SHEN Xiao-qian, ZHAN Ze-li
Location: Yiwu City, Zhejiang Province
Total Area: 9000 sq.m.
Design Time: September, 2014-December, 2015
Co-operation: Architects & Engineers Co., LTD of Southeast University

设计构思：

　　义乌细菌战受难同胞纪念馆位于浙江省义乌市和平公园内。纪念馆西侧为林山寺，寺内有侵华日军细菌战活体解剖室遗址。建筑主体采用浅色和深灰色两个体块，一个低伏一个高昂，寓意痛苦中的挣扎和生与死的主题。参观内容及主题通过连续的参观流线逐级展开并组织，使得参观一气呵成。

北京亦庄环渤海高端总部基地规划设计方案
YIZHUANG BOHAI HIGH-END HEADQUARTERS BLOCK PLANNING, BEIJING

设 计 人：齐康、张四维、何柯、盛启寰、赵茜、顾媛媛
工程地点：北京市
工程规模：214180 平方米（用地面积）
设计时间：2013
合作单位：东南大学建筑设计研究院有限公司

Designer: Qi Kang, Zhang Si-wei, Gong Cong, Wang Zhou, Zhang Sheng-ya
Location: Xiamen, Fujian Province
Total Area: 214180 sq.m. (Land Area)
Design Time: 2013
Co-operate: Architects & Engineers Co., LTD of Southeast University

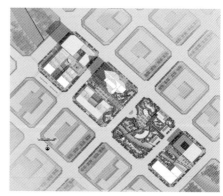

南通大学附属医院新门诊楼设计方案
OUTPATIENT BUILDING OF AFFILIATED HOSPITAL OF NANTONG UNIVERSITY

设 计 人：齐康、王彦辉、蒋澍、宫聪、马程、施鹏骅、朱韵卉	Designer: QI kang, WANG Yan-hui, JIANG Shu, GONG Cong, MA Cheng, SHI Peng-hua, ZHU Yun-hui
工程地点：江苏省南通市	Location: Nantong City, Jiangsu Province
工程规模：27800平方米	Total Area: 27800 sq.m.
设计时间：2014—2016	Design Time: 2014-2016
建设时间：2016.10-	Completion: 2016.10-
合作单位：东南大学建筑设计研究院有限公司	Co-operate: Architectural & Engineers Co., LTD of Southeast University

设计构思：

由于门诊楼建筑具有较强的功能特殊性，而本项目用地狭窄，且周边有西寺等传统建筑群体和需保护的百年银杏树等限制条件，为新门诊楼设计带来较大挑战。

新门诊楼的方案设计，首先立足于满足医院的新发展需求，打造充分体现现代医学理念、功能配置完善、流线组织合理、空间环境舒适的新门诊楼建筑。同时，塑造与濠河景观带及周边环境相协调的建筑形象，使新门诊楼建筑成为地段建筑景观的新亮点。

建筑体块为了尽量减少对濠河景观带及周边环境的影响，方案的处理手法是充分利用场地，使得门诊楼的大厅得到最大程度的开敞，体量上裙房采用层层退后的策略，其中靠近西寺部分为两层，向东依次增为三、四层。屋顶处理上结合屋顶花园做后退处理，以取得与西寺在造型和体量上的呼应。

在立面造型的处理上，以相对纯粹的造型和灰色（墙体浅灰、屋面深灰）色调来使之达到同时与西寺和周边其他风格建筑的协调效果。建筑外墙采用干挂石材、金属翅片与LOW-E玻璃等组成双层建筑表皮，在塑造出简洁现代的新医院建筑形象的同时提升建筑的生态节能效果。

方案鸟瞰图

天台安厦度假酒店设计方案
TIANTAI ANSHA RESORT HOTEL, TAIZHOU

设 计 人：齐康、王彦辉、盛启寰、杨滠葳、于菲、沈骁茜、张涵	Designer: QI Kang, WANG Yan-hui, SHENG Qi-huan, YANG Yi-wei, YU Fei, SHEN Xiao-qian, ZHANG Han
工程地点：浙江省台州市	Location: Taizhou City, Zhejiang Province
工程规模：148560平方米	Total Area: 148560 sq.m.
设计时间：2013–2014	Design Time: 2013-2014

设计构思：

从天台山丰厚文化的资源中汲取灵感与设计依据，将东方传统文化与现代养生休闲相结合，凸显天人合一、道法自然的传统文化精髓，在功能配置、主题打造、建筑环境设计等各个层面予以体现。

高度关注建筑与环境互济共生关系的营造，顺应地形，灵活布局，借鉴当地传统民居街巷结构，空间层次丰富，尺度宜人，步移景异。将原生态自然环境及地方建筑文化与最新设计手法及建筑科技相结合，在传承基础上进行创新，探索具有地方文化特色的新建筑风格。提取地方建筑元素及文化元素，在心理环境、总体布局、建筑布局、建筑内外形态等方面延续并发展浙东低区建筑特色。摒弃对单纯现代、华丽或复古幽情的过度追求，用尽量少的地方元素（如砖、石、木、水等）来营造兼具时代气息和传统韵味的建筑形象及空间意境。

南通东城涌鑫广场设计方案
DONGCHENG YONGXIN PLAZA, NANTONG

设 计 人：齐康、王彦辉、何柯、
　　　　　高晓明、仲早立、葛晓峰
工程地点：江苏省南通市
工程规模：154967 平方米
设计时间：2012–2013
建成时间：2017

Designer: QI Kang,
WANG Yan-hui, HE Ke,
GAO Xiao-ming, ZHONG Zao-li,
GE Xiao-feng
Location: Nantong City, Jiangsu Province
Total Area: 154967 sq.m.
Design Time: 2012-2013
Completion: 2017

设计构思：

设计秉承功能复合与带动周边相结合、标志形象与精神彰显相结合、便捷高效与以人为本相结合、资源集约与生态环保相结合的理念，合理布局特色商业、商务办公、宾馆饭店、旅游服务、餐饮休闲等功能区域，形成功能齐全、空间舒适、环境优美的现代化服务综合体。同时，为了充分满足汽车客运及项目自身未来发展的需求，广场地下与客运站、规划中的城市轨道交通相互连通。

顺应该片区及周边城市空间建设发展趋势、与南通市汽车枢纽东站工程在功能上相配套、形象上协调，建设多功能复合的现代化城市综合体。在实现客运东站的配套服务功能的同时服务周边，提升完善周边区域的城市功能档次与品质。与南通客运枢纽东站一起，共同形成南通市发展中新的标志性形象，并带动城市东部区域的跨越式发展。

青岛大学科技研发中心设计方案
THE SCIENCE R & D CENTER OF QINGDAO UNIVERSITY

设 计 人：齐康、金俊、陈小坚、张曼、
　　　　　李菲、常辰
工程地点：山东省青岛市
工程规模：50650平方米
设计时间：2012.12

Designer: QI Kang, JIN Jun, CHEN Xiaojian,
　　　　　ZHANG Man, Li Fei, CHANG Chen
Location: Qingdao City, Shandong Province
Total Area: 50650 sq.m.
Design Time: 2012.12

设计构思：

　　青岛大学科技研发中心坐落在青岛大学正门中轴线尽端，面向入口广场，背靠山体。建筑对称的造型强化了入口广场的中轴线使校区入口更具气势。建筑采取中空的造型，通过建筑看到山体，起到框景效果。青岛大学科技研发中心提取了青岛历史文化元素，采用古典构图方式和拱券意向，传承了城市文脉。设计上呼应青岛市特有的海洋文化，采用曲线的造型，轻盈灵动，反映时代特色。

青岛大学医学教育综合楼设计方案
MULTI-PURPOSED MEDICAL TEACHING BUILDING IN QINGDAO UNIVERSITY

设 计 人：齐康、金俊、陈小坚、张曼、李菲、胡长涓
工程地点：山东省青岛市
工程规模：47398 平方米
设计时间：2013.1

Designer: QI Kang, JIN Jun, CHEN Xiao-jian, ZHANG Man, Li Fei, CHANG Chen
Location: Qingdao City, Shandong Province
Total Area: 47398 sq.m.
Design Time: 2013.1

设计构思：

医学教育综合楼位于青岛大学北部。用三幢既相对独立又相互联系的高层建筑组成。室外空间层次丰富，既尊重环境，又营造宜人的交流场所。高层布局化整为零，高低错落，呼应山岭形状，延续城市文脉，既有石头的粗犷，又有水的柔美。

武夷山"红山文化休闲（荣昌汇）"规划设计方案

PLANNING FOR RONGCHANGHUI IN WUYI MOUNTAIN

设 计 人：齐康、金俊、顾媛媛、黄梅、张曼、
　　　　　刘九三、乔培森、朱韵卉
工程地点：福建省武夷山洋庄
工程规模：20806平方米
设计时间：2014.08

Designer: QI Kang, JIN Jun, Gu Yuan-yuan, Huang Mei, ZHANG Man,
Liu Jiu-san, Qiao Pei-seng, Zhu Yun-hui
Location: Wuyishan Yangzhuang, Fujian Province
Total Area: 20806 sq.m.
Design Time: 2014.08

设计构思：

武夷红山文化休闲荣昌汇位于福建省武夷山洋庄。用地四面环山，景色宜人。本方案从自然环境、人工环境、生态环境等方面来进行创作，充分利用山地的自然优势发扬中国传统文化，建设良好的生态环境。采用枕山而筑、襟水而立的总体布局和水绿相融、气韵相通的建筑风格。

图书在版编目（CIP）数据

齐康及其合作者建筑设计作品集/齐康著. —北京：中国建筑工业出版社，2018.3
ISBN 978-7-112-21669-7

Ⅰ.①齐… Ⅱ.①齐… Ⅲ.①建筑设计－作品集－中国－现代 Ⅳ.①TU206

中国版本图书馆CIP数据核字（2017）第316716号

责任编辑：张 建 张 明
责任校对：李美娜

齐康及其合作者建筑设计作品集
齐康
*
中国建筑工业出版社出版、发行（北京海淀三里河路9号）
各地新华书店、建筑书店经销
北京锋尚制版有限公司制版
北京富诚彩色印刷有限公司印刷
*
开本：880×1230毫米 1/12 印张：24$\frac{2}{3}$ 字数：447千字
2018年3月第一版 2018年3月第一次印刷
定价：200.00元
ISBN 978-7-112-21669-7
（31291）

版权所有 翻印必究
如有印装质量问题，可寄本社退换
（邮政编码100037）